International Federation of Automatic Control

DISTRIBUTED DATABASES
IN REAL-TIME CONTROL

IFAC WORKSHOP SERIES

Editor-in-Chief
Pieter Eykhoff, University of Technology, NL-5600 MB Eindhoven,
The Netherlands

CHESTNUT *et al.*: International Conflict Resolution Using System Engineering   *(1990, No. 1)*
SIGUERDIDJANE & BERNHARD: Control Applications of Nonlinear Programming and Optimization
    (1990, No. 2)
VILLA & MURARI: Decisional Structures in Automated Manufacturing   *(1990, No. 3)*
RODD *et al.*: Artificial Intelligence in Real Time Control   *(1990, No. 4)*
NARITA & MOTUS: Distributed Computer Control Systems (DCCS'89)   *(1990, No. 5)*
KNUTH & RODD: Distributed Databases in Real Time Control   *(1990, No. 6)*
LOTOTSKY: Evaluation of Adaptive Control Strategies in Industrial Applications   *(1990, No. 7)*
MEYER: Real Time Programming   (1990, No. 8)
MOWLE: Experience with the Management of Software Products   *(1990, No.9)*
TAKAMATSU & O'SHIMA: Production Control in Process Industry   *(1990, No. 10)*

RODD: Distributed Computer Control Systems   *(1989)*
CRESPO & DE LA PUENTE: Real Time Programming   *(1989)*
McAVOY: Model Based Process Control   *(1989)*
RODD & SUSKI: Artificial Intellligence in Real Time Control   *(1989)*
BOULLART *et al.*: Industrial Process Control Systems   *(1989)*
SOMMER: Applied Measurements in Mineral and Metallurgical Processing   *(1989)*
GOODWIN: Robust Adaptive Control   *(1989)*
MILOVANOVIC & ELZER: Experience with the Management of Software Projects   *(1989)*
GENSER *et al.*: Safety of Computer Control Systems (SAFECOMP'89)   *(1989)*

Other IFAC Publications

AUTOMATICA
the journal of IFAC, the International Federation of Automatic Control
Editor-in-Chief: G. S. Axelby, 211 Coronet Drive, North Linthicum,
Maryland 21090, USA

IFAC SYMPOSIA SERIES
Editor-in-Chief: Janos Gertler, Department of Electrical Engineering,
George Mason University, Fairfax, Virginia 22030, USA

Full list of IFAC Publications appears at the end of this volume

NOTICE TO READERS

If your library is not already a standing/continuation order customer or subscriber to this series, may we recommend that you place a standing/
continuation or subscription order to receive immediately upon publication all new volumes. Should you find that these volumes no longer serve your
needs your order can be cancelled at any time without notice.

Copies of all previously published volumes are available. A fully descriptive catalogue will be gladly sent on request.

ROBERT MAXWELL
Publisher

DISTRIBUTED DATABASES IN REAL-TIME CONTROL

Proceedings of the IFAC/IFIP Workshop,
Budapest, Hungary, 16–18 October 1989

Edited by

E. KNUTH

Computer and Automation Institute,
Hungarian Academy of Sciences, Budapest, Hungary

and

M. G. RODD

Institute for Industrial Information Technology,
University of Wales, Swansea, UK

Published for the

INTERNATIONAL FEDERATION OF AUTOMATIC CONTROL

by

PERGAMON PRESS
Member of Maxwell Macmillan Pergamon Publishing Corporation
OXFORD · NEW YORK · BEIJING · FRANKFURT
SÃO PAULO · SYDNEY · TOKYO · TORONTO

U.K.	Pergamon Press plc, Headington Hill Hall, Oxford OX3 0BW, England
U.S.A.	Pergamon Press, Inc., Maxwell House, Fairview Park, Elmsford, New York 10523, U.S.A.
PEOPLE'S REPUBLIC OF CHINA	Pergamon Press, Room 4037, Qianmen Hotel, Beijing, People's Republic of China
FEDERAL REPUBLIC OF GERMANY	Pergamon Press GmbH, Hammerweg 6, D-6242 Kronberg, Federal Republic of Germany
BRAZIL	Pergamon Editora Ltda, Rua Eça de Queiros, 346, CEP 04011, Paraiso, São Paulo, Brazil
AUSTRALIA	Pergamon Press Australia Pty Ltd., P.O. Box 544, Potts Point, N.S.W. 2011, Australia
JAPAN	Pergamon Press, 5th Floor, Matsuoka Central Building, 1-7-1 Nishishinjuku, Shinjuku-ku, Tokyo 160, Japan
CANADA	Pergamon Press Canada Ltd., Suite No. 271, 253 College Street, Toronto, Ontario, Canada M5T 1R5

First edition 1990

Library of Congress Cataloging in Publication Data

Distributed databases in real-time control: proceedings of the
IFAC/IFIP Workshop, Budapest, Hungary, 16–18 October 1989/edited by
E. Knuth and M.G. Rodd.—1st ed.
    p.   cm.—(IFAC workshop series: 1990, no. 6)
"IFAC Workshop on Distributed Databases in Real-Time Control,
organised by Computer and Automation Institute, Hungarian Academy of
Sciences, sponsored by the International Federation of Automatic
Control, Technical Committee on Computers, co-sponsored by
International Federation for Information Processing TC5, Computer
Applications in Technology".
Includes index.
1. Distributed data bases—Congresses.   2. Real-time control—
Congresses.   I. Knuth, E. (Elöd), 1943- .   II. Rodd, M. G.
III. International Federation of Automatic Control.   IV. IFAC
Workshop on Distributed Databases in Real-Time Control (1989:
Budapest, Hungary)   V. Magyar Tudományos Akadémia.
Számitástechnikai és Automatizálási Kutatóintézet.
VI. International Federation of Automatic Control. Technical
Committee on Computers.   VII. IFIP Technical Committee 5—Computer
Applications in Technology.   VIII. Series.
QA76.9.D3D583 1990   005.75'8—dc20   90–7277

British Library Cataloguing in Publication Data

Distributed databases in real-time control.
1. Distributed digital control systems
I. Knuth, Elod   II. Rodd, M. G.   III. International
Federation of Automatic Control   IV. Series
629.895
ISBN 0–08–040504–5

These proceedings were reproduced by means of the photo-offset process using the manuscripts supplied by the authors of the different papers. The manuscripts have been typed using different typewriters and typefaces. The lay-out, figures and tables of some papers did not agree completely with the standard requirements: consequently the reproduction does not display complete uniformity. To ensure rapid publication this discrepancy could not be changed: nor could the English be checked completely. Therefore, the readers are asked to excuse any deficiencies of this publication which may be due to the above mentioned reasons.

The Editors

This title is also published in *Annual Review in Automatic Programming*, Volume 15, Part I.

ISBN   9780080405049

Printed in Great Britain by
CPI Antony Rowe, Chippenham and Eastbourne

IFAC WORKSHOP ON DISTRIBUTED DATABASES IN REAL-TIME CONTROL

Sponsored by
The International Federation of Automatic Control:
Technical Committee on Computers

Co-sponsor
International Federation for Information Processing
TC5 — Computer Applications in Technology

Organized by
Computer and Automation Institute, Hungarian Academy of Sciences

IFAC WORKSHOP ON DISTRIBUTED DATABASES IN REAL-TIME CONTROL

Sponsored by

The International Federation of Automatic Control,
Technical Committee on Computers

Co-sponsor

International Federation for Information Processing
TC-5 — Computer Applications in Technology

Organized by

Computer and Automation Institute, Hungarian Academy of Sciences

International Programme Committee

M.G. Rodd, UK (Chairman)	J. Kramer, UK
D. Beech, USA	K. A. Meerman, Netherlands
B. Eriksson, Sweden	T. Peng, PRC
W. Ehrenberger, FRG	F. L. de la Puente, Spain
P. L. Elzer, FRG	A. A. Stogny, USSR
T. Unger, USA	W. W. Suh, Korea
V. H. Haase, Austria	D. Tolhoek, USA
D. Hutchison, UK	A. B. Whinston, USA
L. A. Kalinichenko, USSR	

National Organizing Committee

E. Knuth (Chairman)
L. Sa. Berényi
J. Harangozó
G. Kovács
J. Szlanko

PREFACE

Despite the long history of the component disciplines (such as database management and communication standards) distributed databases in real–time control cover fields where brand–new technologies are needed. While in other environments the sophisticated high–level conceptual layers and other complex conceptual structures are vital to ensure system integration or to make the complexity manageable, here they are useless.

The traditional high–level conceptual models, therefore, are generally not applicable here (more precisely, others are more desirable); on the other hand, the huge transaction fluxes involved, and the rigidity of the time constraints, prohibit the use of conventional solutions as well.

Presently, in this field, we cannot speak about quasi–standard solutions at international levels. However, efficiently applied systems exist, realised mostly on an *ad hoc* basis. Such case–studies are of vital importance for the future and I am convinced that these are going to lead to further pre–standard solutions – and this volume contains papers on many such undertakings.

Also, I do not share the opinion that existing theories cannot provide basic help when realising distributed real–time databases. In the fields of database–management, communication and process control, vast knowledge is at our disposal, providing – if nothing else – at least the limits to the possible paths which may be followed.

During this Workshop, representatives of both extremes met. Those wishing to solve industrial problems at whatever cost were sitting alongside the academic representatives of specific disciplines and with the dedicated advocates of international standards. All in all, it proved to be a successful combination.

Elod Knuth

CONTENTS

Contents

REAL-TIME ISSUES IN DISTRIBUTED DATA BASES FOR REAL-TIME CONTROL

M. G. Rodd

Institute for Industrial Information Technology, University of Wales, Swansea, UK

Abstract: This paper takes a pragmatic application-oriented view of distributed data bases to be used in Real-Time Process and Manufacturing Control. It acknowledges the strides that have been made in the development of distributed data base techniques in general, but is concerned with the problem of mapping these techniques into the Real-Time Control world. The paper takes the view that distributing data in a control system is not only an essential feature of integrated systems, but indeed is the key to their success or otherwise. However, this distribution of data has characteristics which differentiate it from other areas in which distributed data bases are used - such as air line bookings, etc. The paper discusses the essential nature of these differences and goes on to discuss how distributed data bases have to fit in with the overall control strategy, which in turn calls heavily on the use of integrated communication systems. These latter systems are themselves often based on rapidly emerging standardised systems (primarily for economic reasons), and are increasingly OSI-based. The paper concludes that to be of value, a distributed data base in a real-time control system must be designed from the viewpoint of the user rather than from that of the data base designer!

Keywords: Real-Time Distributed Data Bases, Real-Time Issues in Data Bases, Distributed Data Bases, Distributed Real-Time Data Bases, Integrated Data Bases.

1 TOWARDS INTEGRATED CONTROL SYSTEMS

In all areas of automation, there is a very natural, evolutionary trend towards total integration of the controlling structures. The reasons for this are very simple: economics! Clearly, with the competition which has grown up in the last 20 years to produce products which are cheaper and more reliable, and simply satisfy the market need, has required manufacturers of any product to look towards achieving total efficiency, be it in a chemical process or a discrete manufacturing plant. The key to such efficiency is to have total control. Fundamental to total control is data.

Data is now seen as a valuable resource. Just as a local control system cannot operate without knowledge of the process it is controlling, or without a set point provided from a superior system, in a similar way the managing director of a company cannot make correct policy decisions without knowing what is happening on his plant at any one time, as well as what the market needs are.

Automation is naturally not the panacea to these problems; indeed, a small, tight, manufacturing plant, run by a manager who is totally in control of that plant, and has reliable knowledge of the market demands, can be extremely efficient. However, as our plants get more complex in terms of the interactions between their various components, and indeed, as the relationships between the plants and the suppliers get more and more complicated, the only solution is to turn towards mechanical and electronic aids to achieve the integration. In non-philosophical terms, integrated computer control is simply a means by which a complex system, with high-speed responses at all stages, can be effectively tied together.

As a result, distributed computer control systems are neither magic nor anything more innovative than a natural solution to a very tough problem - a problem essentially of coping with complexity. We have learnt only too well the problems which arise in producing complex systems. Not only do we have difficulty in conceptualising how they should be structured but, more important, we have come to realise our own fallibility. We simply cannot manufacture either hardware or software with the required degree of reliability. As a result, we learn very quickly that the only way of tackling this problem is that of "divide and conquer". We are rapidly learning to divide up complex software tasks into relatively small chunks - chunks which we can understand and monitor, and at least attempt to test.

As we move, then, towards distributed computer control structures, it has become evident that we must produce some form of hierarchy in order to structure these systems. We have to harness geographically related processes and wherever possible reflect the natural integration which results. We also find that as we develop an appropriate form of hierarchy in our distributed computer control systems, the various levels of automation define the information requirements. In practice, the so-called five-layer model, as illustrated in Figure 1, seems to become relevant, reflecting as it does most practical situations.

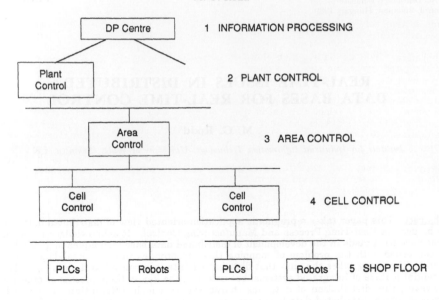

Figure 1

A further requirement of the distribution is to cope with the problem of system failure. During the early days of computerisation we learned, to our cost, the disasters which result from the "eggs in one basket" syndrome. The loss of an essential computer could be devastating. We have learned that the only way of tackling this is to distribute the control, so that the failure of any particular piece of software or hardware will lead to a situation which we can at least contain.

Naturally, all that has been discussed so far is very evident and the points do not need labouring. However, it is important to realise that fundamental to such integrated systems and their hierarchical implementation is the data which is the "lifeblood" of any distributed system. Equally important is the fact that the "veins" which carry this lifeblood are the communication channels which we instal between our computing nodes. As we will see later, the nature of these "veins" has a tremendous influence on how we move information around the plant or, in a grander way, between plants using various public services.

Turning to the question of data, our colleagues in the data processing industry will say that there is no problem, and that most of the difficulties were resolved several years ago. They point, very rightly, to the successful airline booking systems, or to the international banking structures, all of which naturally demand distributed data bases. However, when one starts mapping these techniques into our real-time control system applications, various things appear to go wrong, and there does appear to be a fundamental difference between the very successful applications of distributed data bases in the data processing field and those which are encountered on the shop floor.

2   REAL-TIME VERSUS ON-LINE

The thesis of this paper is that to design a real-time control system one must appreciate the real-time nature of the process that one is controlling. In essence the problem is one of distinguishing between an "on-line" computing system and a "real-time" computing system. This distinction has become extremely blurred over many years, and it is critical that we return to some fundamental understanding.

Essentially, in a real-time computing system the correctness of the system depends not only on the logical results of the computation but also on the time at which the results are produced. "Time-critical" or "real-time" does not necessarily imply "fast", but does mean "fast enough for the chosen application".

In essence, in time-dependent processes, data is judged not only on its actual value but also on the time at which the value was obtained. For example, a sensor measuring an attribute of a dynamic system should specify not only the actual value of that attribute but also when that parameter was sampled, since of course the system is time-varying. Real-time data in any distributed network is thus valid only for a given period of time and indeed has meaning only if the time of creation is specified. This forces designers increasingly towards the concept of atomic units of data in which the value and the time of creation are inseparable. Fundamental to all this is that we are completing control loops. The designer of a simple PID controller assumes that the variables which we are measuring are obtained when we think they are. However, if we put a transmission system between the point of measurement and the point at which the measurement is utilised, that transmission system has its own characteristics. These characteristics must be considered in solving the control algorithm problem. There is no point in comparing the value of a temperature sensor with a required set point unless we know at what point in time we are actually measuring that temperature. On the other hand, in the on-line situation, we are merely required to be able to have discussions directly with our computer. If that computer takes a little bit longer to handle, say, a request, then this is not serious, except for maybe causing some irritation.

In the real-time system, any delay could cause the plant to be in serious error.

It is clear, therefore, that the data which flows around our distributed computer control system will, in many cases, be time-dependent and not merely value-dependent. It is important, therefore, that we look at the nature of the data in our real-time control system when we start developing methods of transmitting and storing that data as the links in integrated computing systems, i.e. the sharing of data.

Professor Hermann Kopetz of the Technical University, Vienna, proposed some years ago in a public lecture that the characteristics shown in Figure 2 give us a good clue as to the difference between "real-time" and "on-line". He and many other workers in the area have suggested that there are various degrees between these two extremes. However, it is felt in this paper that this distinction is not really necessary and that a simpler and more reasonable approach would be to define a real-time process as one in which the temporal characteristics are absolutely defined, whereas an on-line situation is one in which these can be degraded without any serious loss of performance.

As a result, it is suggested that in real-time control, data has meaning only when it is associated with time. This implies, in the first place, that all data must be time-stamped at the time of creation in order to provide that real-time information. Of course, the direct consequence of this is that all nodes in a system must have access to a globally agreed real-time clock. Again, work by many authors has pointed to ways in which such clocks can be created. In essence they require a good-quality local clock at each computing node, with some form of synchronisation by means of frequently transmitted time messages, or by a separate timing channel, or, finally, by access to some internationally transmitted standard.

However, merely time-stamping the data does not solve all the problems. A particularly important issue which must be considered, and which will be referred to later, is that of consistency. Not only must data be consistent in value, but this consistency must be referred back to time. Two aspects of this are well illustrated by the following examples.

In the first place, consider the implementation of redundant sensors. Say, for example, one is measuring a single temperature variable but using three redundant temperature probes, with each probe producing its version of the temperature. A further controller must make use of this temperature information, and the question of course arises as to which temperature value to use. Whilst in the past this would probably have been handled by a voting mechanism, clearly if each probe transmits its information separately over the communication network, then there is no way for the controlling node to know how to implement this voting, since it will not know when each value has been created. The only way to cope with the problem is by time-stamping those temperature variables.

The second example is a very complex one, relating to the problem of an operator getting a consistent picture of an overall plant. Consider that the operator might be at least two or three networks away from the actual plant; then information which is presented to him on the screen will have to travel up the communication networks before it can be displayed. Information which is obtained locally could well get to him before information which is obtained from a remote site. If, however, he (or an accompanying artificial intelligence system) is required to base a decision upon the data presented, this is an impossibility unless the actual time at which all the data was created is available to him, thus ensuring that consistency can be checked.

Both these points have great bearing on the structure of the distributed data base, and this will be referred to later in this paper.

REAL-TIME versus ON-LINE

CHARACTERISTIC	HARD REAL-TIME	ON-LINE
Response time	Hard	Soft
Pacing	By Plant	By Computer
Peak Performance	Must be Predictable	Degradable
Time Granularity	Less than 1 mSec	About 1 Sec
Clustering	Important	Not Essential
Data Files	Small-to-Medium	Large
Data Integrity	Short Term	Long Term
Safety	Critical	Non-Critical
Error Detection	Bounded by System	User Responsible
Redundancy	Active	Standby

Figure 2

It is also relevant at this stage to point out that when one looks very closely at the hierarchical structure which is developing in a distributed computer control system, one sees that the types of data which occur at the various stages in the process vary greatly. This, again, will be referred to later in this paper.

3   COMMUNICATION SYSTEMS IN DCCSs

So far we have mentioned the fact that in a Real-Time Distributed Computer Control system it is critical to take into consideration time and its effect on data.   Supporting our distributed computer system in the real-time control world, of course, will be communication systems, and their very nature will have a great influence on what is moved around the plant, and how. We have too often been guilty of assuming that because we appear to have very large bandwidths available to us in our networking systems, we are free to move as much data as we wish around a plant. In reality, however, once we try to move this data reliably and efficiently we suddenly discover that our communication systems start to become bottlenecks in themselves. Also, we suddenly discover that the nature of the communication systems can greatly influence the efficiency of moving the data. Particularly relevant, too, is that once we realise the real-time nature of data in process control then we have to look at the real-time characteristics of our communication systems.

Of course the fundamental issue in the development of any interconnected computer structure is the problem of incompatibility.   Essentially, in a distributed computer system what we are trying to do is to permit various computers from various manufacturers to share various pieces of data. Whilst from the point of view of the supplier it would be desirable for all our computers to be from his own stable, and therefore totally compatible, we soon realise that in the distributed computer control world we are at the mercy of a variety of suppliers, each of whom has particular strongpoints; thus, certain suppliers can provide very fine data processing hardware and software, whereas others specialise in hard-nosed factory floor compatible controllers. The difficulty of producing a distributed system is to bring all these bits and pieces together. The compatibility issue revolves around hardware, software and, of course, data. Of course, we would like to be free to design our own communication systems and indeed our own programs.   However, the reality is somewhat different!

In practice, solving the compatibility problem is extremely difficult and we are only beginning to make some form of impact on approaching it. The key has to be standardisation - whether or not we or the suppliers will acknowledge it. Naturally, at the upper data communication level much progress has been achieved of late in the move towards OSI-based protocols, and in other fields, too, the success of standardisation is evident - for example Ethernet and TCP-IP are good examples of standards which have met their desired marks.

Of course in the hard world of process control standardisation is critical, but unfortunately we are relatively small as an industry, when compared to the data communications field. Whilst we would like to set our own standards there is no way that we can economically go down this route and therefore we tend to be the "step-children" of the data communications industry.

Realising the problems of standardisation, General Motors initiated its MAP exercise some years ago and at the same time Boeing Corporation developed its TOP initiative. In both cases the idea was to produce a profile of protocols which would be appropriate for their particular areas of application - in the case of MAP, manufacturing, and in the case of TOP, technical offices. It was realised right from the start that there was no point in going it alone, and that the prime way ahead would rather be to select from an agreed and internationally acceptable protocol profile the appropriate layers, and also within those various layers to select certain protocols and, where necessary, to inject extra emphasis to direct the move towards standardisation.

Although MAP has attracted much criticism, it has undoubtedly been very significant in that it has brought to the fore the need for standardisation. The exercise has also hastened the development of protocols which are appropriate (at least to some degree) to manufacturing and to the process industry. It is very clear, however, that the products which are resulting from this exercise are in many cases too expensive and too complicated to be of value in the hard world of automation. However, at the same time it must be acknowledged that certain aspects of the MAP exercise will undoubtedly prove to be extremely important in the long term.

Of particular importance, for example, is the development through MAP of MMS - Manufacturing Messaging Services. MMS provides a standardised manner of sending messages around a process or manufacturing plant. The standard itself defines a core of services, and then it is left to specialised groups to provide so-called "companion standards", which expand on these services for particular application areas. Indeed, it is probably fair to say that the standardisation exercise should have started at the MMS level, leaving many of the lower-layer and somewhat technical issues for later, in that many of the lower layers, such as the physical layer, will be highly influenced by technical achievements and current issues (for example, whether it should be Ethernet or Token-passing) might well eventually be rendered redundant.

The point is that there can be little doubt that the standardisation exercises are extremely critical to the design of future distributed computer control systems.   They will also have serious impacts on how we handle data within DCCSs. We cannot ignore them: they are economically critical and simply to throw out the MAP exercise would be extremely naive. However, to look at the MAP exercise, to take what is good out of it and then to assist in the development or redevelopment of some of the layers which are proving to be troublesome, makes eminent sense.

Another aspect of the trend towards standardisation fits in extremely closely with the idea of development hierarchies within the integrated control system. Exercises such as MAP have acknowledged that we will ultimately see, say, a high-capacity data backbone running through the plant, fed by bridges or gateways from less-capable, cheaper networks, which in turn could well be fed by very low-cost field-bus type networks. After much debate and academic discussion has taken place as to the ideal structure there does appear to be an increasing consensus that our systems will ultimately move, in the case of real-time Control Systems, to a structure like that illustrated in Figure 3.

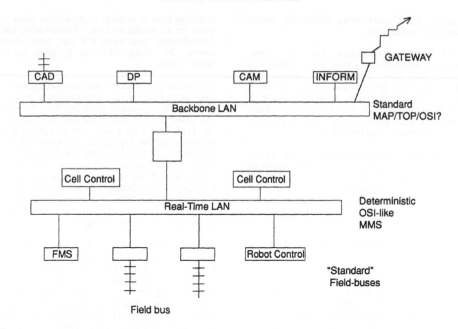

Figure 3

In this figure one sees, again, the idea of having a powerful backbone LAN which can in turn talk via gateways to wide area networks. The backbone will also communicate into more deterministic high-speed networks which we have chosen to call real-time LANs. Whilst it is clear that current OSI proposals do not cope with the requirements of such real-time systems, there is no doubt that in the future they will. It is interesting to note also that many important companies operating in this area are envisaging models very similar to that proposed in Figure 3. It is critical also to note that below the real-time LAN level there will be a move towards so-called Standard Field-Buses and here the current international attempts to standardise the field-bus are particularly relevant.

Important throughout this is that whilst various components are not yet defined through International Standards, there is no doubt that in time they will be. More important, though, is the acceptance of the OSI model; probably even more important, is the acceptance of some of the higher-level services such as MMS, FTAM, Directory Services, etc.

4   LEVELS AND FUNCTIONS

Studying Figure 3 in detail and looking at the data requirements at various levels one begins to see a pattern emerging.

At the higher levels, such as the backbone LAN, one sees that the functions which are typically required include

- File transfer
- Electronic mail
- Remote file editing
- Status reporting to operators in higher levels of management
- Data acquisition from lower levels
- Control of lower levels
- Program transfer

In essence, at this level, one is moving a large bulk of data around, typically between relatively large machines. This could be between CAD systems, or by sending plant status information through to management computers running production scheduling or manufacturing materials require-ment planning. At this level one can see the need for efficient, but safe, transfer of bulk data. However, if one is additionally utilising this information for decision-making, then it is necessary also to be able to send time-tagged information across this system, otherwise no useful decisions can be made! However, one can accept at this level of networking that delays will occur, particularly when large files are being transferred. It is clear also that one could almost consider this an on-line environment, provided that there is some form of access to real-time data where necessary. Moving to the lower levels, particularly at the real-time LAN level, and in fact at the field-bus level, one sees a slightly different collection of functions. Here one is really getting into the real-time control world, and the following characteristics start to emerge

- Minimal defined message delay times
- Deterministic behaviour - particularly under crisis situations
- Real-time synchronisation
- Assured data consistency
- Inherent message redundancy
- State broadcasting
- Regular and frequent data updating

Essentially, we are talking here about highly efficient data transmission. Speed is important, but only to the extent that it must match the response of the plant. Thus a very slow plant will not require very high transmission rates. However, what it will require are deterministic rates. Information which flows at this level is typically relatively small in quantity, but critical in terms of its determinism. For instance, if a program does not complete on time, this could lead to serious error conditions. It is important also to realise at this stage that the network system must work most efficiently under the worst possible conditions.

Looking at our networking (or, in fact, internetworking) strategies in this way, gives us a clue as to how they may best be implemented, and what technologies are appropriate at which level. That there are different functions cannot be doubted, but a fundamental criterion has to be that where data is associated with time then the communication systems must inherently support this.

It is important also to emphasise that we cannot ignore international standardisation. As much as we would like to reject many of the proposed standards, for economic reasons we simply cannot. What we have to do is work with the standards, and alongside them, in striving towards systems which do fully meet our real-time requirements.

5   DATA BASES IN DISTRIBUTED COMPUTER CONTROL SYSTEMS

Finally, we look now at the problem of distributing our data bases in a real-time control system. Again, we must point out that the viewpoint taken in this paper is that the real-time control system starts at the highest level and works all the way through, right down to the plant regime. Distributing our data is, therefore, a very natural function, not merely to ensure security but to assist in meeting our time objectives. It is extremely important to examine why we move data about a plant and this could well be the key to the design of the appropriate data base structure.

The point is that at the lowest level we might just be measuring a simple temperature variable and using this to control the output of a temperature controller. At the higher level though, this information is not required. What is required is the knowledge that the controller is functioning and keeping its temperature to a specific value.

This analogy can be carried on throughout the process. The Managing Director of a company might merely want a simple spread sheet which contains information which can allow him to make appropriate decisions. He doesn't give a damn about the value of the temperature setting! What he does give a damn about, however, is the fact that the process which uses the sensor is operating, and what the actual production rate is. He needs a complete picture of what is happening on the plant, but a highly refined one.

Thus, the refining of data becomes extremely important. The simple principle should always be used - don't transmit anything unless it is needed at the other end! Thus, whilst we have regarded our communication systems as having virtually limitless band-width, in fact this is far from the truth and we are learning at our cost that once we provide effective and reliable protocols, they impose such loads on our communication systems that a 10Mbit per second communication system suddenly has an effective band-width of a handful of messages per second!

The guideline, therefore, to the distribution of our data bases has to be that we maintain the data where it is actually required. Data must be seen as a critical resource, and must be made available only where, and when, it is actually required. This in turn requires us to know, at a very early stage, who is going to use what, and when. Again, if the user

requires access to data in real-time then this facility must be provided to him. Essentially, then, we are distributing the data for the convenience of the users, be they Managing Directors or local controllers.

In practice, it is found that the actual size of the data base is in itself a function of where in the hierarchy it sits. Experience is showing that the data bases at the lower levels, which might well be referred to as the real-time data bases, are relatively small in size, whereas at the higher levels where the data bases reflect the global situation, we end up with extremely large data bases. Of course it is also common experience that in many processes at least three quarters of all data gathered is collected on a Write Only basis, i.e never again used! (This is probably more a reflection on our inability to specify our requirements fully, though.)

Key, too, to the design of our data base structure is the question of redundancy. There can be no doubt that as our plants become more and more complex, we must distribute the data, just as we are distributing the hardware and software to cope with the problem. We have to make certain, therefore, that we can replicate data bases in such a way that if we lose either a data base or the associated computer, we can either bring in new hardware or execute programs on existing hardware.

This latter problem is effectively the same as is met whenever we want to copy any information to more than one data base - the problem of data consistency.

Data consistency in real-time data bases is of fundamental importance and must be taken into consideration at the earliest stages of any data base design. The problem is naturally only too well known in the distributed data base community in on-line situations, but when we start to apply these techniques in the real-time environment the situation changes. Logical techniques are no longer of any value. The only techniques which stand a chance of working are those related to real-time. Data base consistency in the real-time environment can only be ensured by taking into consideration the real-time nature of the data and the absolute binding which occurs between value and time. Therefore, solving the consistency problem can only be effectively undertaken by considering the real-time nature of the data.

Another critical characteristic of the design of real-time distributed data bases has to be the real-time nature of the process to which they are being applied. If one is looking at the lowest level where one is using the data actually to control a real-time control loop, then it is very clear that the access to the data base must itself be totally deterministic. So, just as we require determinism in the communication system which has been discussed previously, we must also strive towards totally deterministic data bases - particularly at the lower levels.

At the higher levels, whilst the access times themselves do not have to be deterministic, we have to cope with the problem of consistency. Here the point is that we are providing information to an operator, or at least to someone who is having to make a decision based on global performance of the system under their control. In order to do this we have to present them with a completely consistent picture of the plant. As we have stressed in this paper, the actual value of any particular variable has significance only if it is associated with the time at which it is created. Therefore, to provide an operator with a consistent picture, this component of the information must be considered.

6   CONCLUSION

This review paper has stressed the fact that in a real-time control system we cannot dissociate the value of data and its time of creation. In the design of a distributed data base we must take this as a prime fundamental consideration. We have to accept that a distributed computer control system is indeed controlling a process which has temporal characteristics. The control system, therefore (be it at the local controller level or at the larger managerial level), must take into consideration this very simple fact of life. Data base consistency is therefore fundamental and can only be achieved with an eye on the real-time nature of the data. Decisions have to be made according to a temporally consistent picture.

The paper has also mentioned the fact that moving the data around is itself critical; not only to the overall strategy but also indeed to the data base, since it will affect the temporal correctness or otherwise of the data. Therefore, in designing distributed computer control systems the fundamental nature of the communication system is itself critical. Also, in designing the data base structure, the nature of the communication system which is going to feed information into it must be taken into consideration.

The basic conclusion has to be that we need to look towards the design of real-time Distributed Data Bases from the point of view of the application, and not necessarily base our ideas around those which might have been very successful when applied in environments which are not analogous to our control world.

7   ACKNOWLEDGEMENT

The author wishes to acknowledge the contributions made by his colleagues at the University of Wales in Swansea, particularly Mr Ivan Izikowitz, Mr Keith Baker, Dr Farzin Deravi and Mr Guo Feng Zhao. Also he wishes to acknowledge the input of his IFAC colleagues, Professor Hermann Kopetz of the Technical University, Vienna, Dr Gregory Suski of Lawrence Livermore National Laboratories and Dr Thierry Lalive d'Epinay of Asea Brown Boveri, Europe.

DIGITAL DESIGN WITH A
DISTRIBUTED ENVIRONMENT IN MIND

J. K. M. Moody

University of Cambridge, Computer Laboratory, New Museums Site, Pembroke Street,
Cambridge CB2 3QG, UK

Abstract. Establishing a real-time control system involves interaction between the design phase and the operational phase. In a simple model the former may be regarded as preceding the latter, but in practice operational experience will yield the data that shape an improved design. Database support must cover both phases, and the enterprise model maintained should be directly accessible to the design tools. Distribution of the application implies distribution of the real-time dependent portion of this database at least. In the operational phase there is a choice between transmitting all control input to a central site and transforming a subset of the control variables at each of a number of locations. Centralized control algorithms can be run on more powerful processors, but communication delays may offset any advantage that could be gained. The impact of uncertainties in computer processing on the stability and performance of a control system can be estimated using the structured singular value of the compensated plant model. In a distributed application design tools should exploit such methods to provide a rational basis for deciding how to distribute control.

Keywords. Database management systems; real time computer systems; distributed control; control system design; stability robustness; structured singular value.

INTRODUCTION

The invitation to present a paper raised a variety of responses. The theme of the Workshop stresses the 'real-time' context. My major experience with real data has involved modelling parish records describing life in an English village across four centuries! The problem tackled was one of model extrapolation in a centralized and static database, with no update except for error correction. More recently my interest has shifted to distributed databases, and to the trade-off between consistency and availability in particular. When a database is distributed across a network it is important that communication failure does not block work which could reasonably be carried out. For example, suppose that a transaction submitted at a workstation does not write any non-local data, but needs to read remote data in order to check some integrity constraint. If such a transaction can be certified in some way it may still be possible to schedule its execution during network partition. Whether or not it is sensible to do this will depend on the details of the application.

Although the main motivation is to make replicated data more available during network partition there are other reasons for the study. Performance considerations in a real-time environment may militate against ensuring the consistency of replicated data structures: examples occur in distributed operating system design, for which it may be advantageous to retain potentially obsolete cached data and to detect rather than to prevent such inconsistency.

This latter interest represents my closest point of contact with real-time applications. Initially transaction processing in distributed databases adapted methods developed from centralized systems (Lindsay and others, 1979). Recent work on algorithms for concurrency control and recovery in distributed database systems has produced alternative strategies (see for example Bernstein, Hadzilacos and Goodman, 1987). New approaches were essential, fundamentally because failure of a single central site is total, whereas in a wide-area distributed system it may not be possible even to establish that 'all' nodes are accessible. For example, in the DEC Name Server there is no requirement for changes to be acknowledged by all sites as rapidly as communications will permit, and the aim of the *sweep* algorithm implemented is to ensure that consistent updates are propagated through the network in 'reasonable time' (Lampson, 1986). What *is* 'reasonable time' depends of course on the application, and an algorithm such as this would be quite inappropriate for a general distributed database management system. The DEC Name Server identifies a common phenomenon in distributed applications, that the significant time constant for users may be long relative to the overall update rate of the application. Another example is that of stock control in a retail business with multiple outlets: for the storeman's purpose there is no need to generate a new consistent database state after each till receipt, indeed it is clear that system performance will only be acceptable if updates are batched in some appropriate way.

There are two distinct points at issue here, and both are relevant to real-time control. Firstly, an application may be tolerant of obsolete information. Indeed, no database that reflects real-world change can be fully current: the crucial question is one of degree, namely the extent to which decisions can be based on the data to hand. Secondly, suppose that in a control application values for manipulated variables are to be computed digitally. Executing any algorithm will involve computation time: if data from more than one site is required then there will be an additional communication delay. These ideas lead from the trade-off between availability and consistency to a real-time interpretation relevant to digital control. If algorithms are to be performed at a required frequency on data collected from a real-time process, then there is a limit to their sophistication. For a given processor, calculations are limited by computational complexity. If data from more than one source is required, or if processing is to take place at a more powerful but remote processor, there may be significant network delays. Thus there is a trade-off between the timeliness and the precision with which results can be obtained. Having discovered an apparent connection between my current interests and control theory was a relief, but before trying to develop the idea I consulted a colleague in the control department, Jan Maciejowski. His response was polite, but firm. With a correct design the controller would be robust, and its possible deviation from the ideal should be quantified from the start. My consideration was relevant, but the right approach was to work within the design methodology of control theory - a methodology of which I knew all too little. The only solution was to begin to remedy the deficiency, and I was pointed to Francis(1987) and Maciejowski(1989).

DATABASE SUPPORT FOR CONTROL IN A DISTRIBUTED ENVIRONMENT

I assume in what follows that the plant to be controlled is distributed, with components located at a number of distinct sites. When the plant is in operation statistics relating to performance are collected: the control design may be adjusted in the light of past performance, or in response to economic or environmental factors (Garcia, Prett and Morari, 1989). Ideally there will be an integrated database covering all of these aspects. Figure 1 shows some of the interactions between them. Despite these interactions the design phase and the operational phase have quite different requirements, particularly with respect to location and response.

At the design phase there is nothing in particular to distinguish the database requirements from those in many other Computer Aided Design applications to complex engineering problems. It is likely that technical packages that can resolve specific mathematical, structural, chemical or other questions will be used heavily: it is vital that design data can be reformatted easily for submission to packages implemented in a variety of programming languages. The response requirement is that of a typical software development project, with the proviso that substantial CPU power may be needed to run such technical packages. A standard database transaction mechanism will be needed, but in addition there should be version support for managing changes that may take weeks or even months to complete. Such

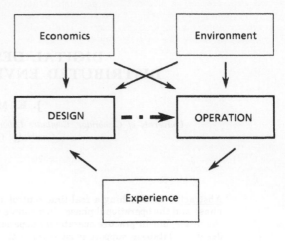

Fig. 1.

version support may also handle the construction and testing or simulation of complete systems that coordinate a number of modifications to the current production system. There is no reason to believe that these activities are inherently distributed whatever the nature of the application, though it is likely that individual design engineers will have powerful personal workstations accessing a common database across a local area network. Data may be replicated, but it is more likely that designers will copy relevant fragments from the common database to their workstation without worrying about subsequent updates: indeed, the version support outlined above should handle this. A recent paper proposes a detailed model for supporting long transactions in just such a design environment, and describes mechanisms for implementing the concurrency controls required (Korth, Kim and Bancilhon, 1988). This transaction model is not tied to a particular model of data, and could be applied in an object-orientated database as readily as in a conventional one. In practice design databases are likely to be complex, involving semantic constraints that are not supported by conventional database management systems (DBMS).

At the operational phase the data handling requirements are quite different. Assuming that control variables are manipulated by digital sampling and real-time computation, the crucial requirement is to keep up with the clock: analysis of the system model will determine what sampling rate is needed, and the structure of the controller will dictate the complexity of the calculations. Although statistics will be gathered and may influence control design and hence future operation, there is no reason to believe that there is *time-critical* interaction with the database during the normal course of events. On the other hand, the abnormal will occur from time to time, and there must be provision for rapid escape from an automatic algorithmic environment to one in which the specific problem can be diagnosed. The ability to combine synchronous event handling involving large quantities of data with priority scheduling is not supported by conventional DBMS. The considerations in the previous paragraph are all concerned with timing only, but for digital control they are likely to force local processing of some data at least. Where high frequency components in the input are expected, or where abnormal behaviour may generate high frequency disturbances, data transmission delays will rule out processing at a central site.

THE ROLE OF DESIGN IN CONTROL APPLICATIONS

The basic methodology of control theory does not seem strange to a computer scientist. The subject has grown up in response to need, motivated by the many complex control problems that arise. Finding practical solutions depends as much on experience as on any precise discipline, and many of the more valuable tools are descriptive rather than analytic. This is true particularly in the multivariable case, since there is often no immediate generalization of single-input single-output techniques. The development of classical control theory has led to descriptions and methods that emphasise response in the frequency domain. A single-input single-output plant can be approximately characterized by its transfer function $G(s)$, and the stability with a given feedback controller $K(s)$ established using Nyquist's criterion, for example.

In the past two decades multivariable control theory has developed rapidly, and many classical results have been extended to transfer matrices $G(s)$, where each element $g_{ij}(s)$ is a transfer function specifying the frequency domain dependence of the ith output variable on the jth input variable. An important result of this period is the 'Youla parametrization', which characterizes all linear stabilizing controllers for any linear plant that can be stabilized. Over the past ten years this technique has led to algorithms for controller design based on a rigorous theory, so called H_∞ optimal control. The basis is a frequency domain theory which depends on the properties of operators on Hardy spaces (Garnett, 1981). In this context systems with time delays or with irrational transfer functions can be modelled (Maciejowski, 1989), though there will not necessarily be a corresponding finite-dimensional state space representation.

The main advantage of the Youla parametrization is that it generates all stabilizing controllers. In practice, however, there may be structural constraints on controller design which do not map naturally to the corresponding Youla parameter. In such a case it may be more appropriate to parametrize controllers in a way that guarantees the structure, and then to solve an optimization problem subject to the explicit constraint of stable control. Provided that a suitably structured approximate controller can be derived and that control objectives can be quantified, an acceptable design should be reached; two quite different approaches are presented in Edmunds (1979) and Kreisselmeier and Steinhauser (1983). In particular, if a distributed application lends itself to a number of local control loops together with central coordination this can be modelled by a controller having an explicit structure. Optimization within this structural constraint will not of course lead in general to a truly optimal controller, but tools exist to analyse the robustness of a controller subject to such constraints. Doyle (1982) introduced the structured singular value $\| Q \|_\mu$ for a compensated plant model under structured perturbations, and derived many of its properties. This measure is used in Doyle, Wall and Stein (1982) to analyse the robustness of performance and stability for such a compensated plant. These perturbations model specific uncertainties in the plant or the controller model, and are constrained to have block diagonal structure overall. The measure $\| Q \|_\mu$ (in fact not a true norm) is defined in such a way as to exhibit a natural dependence on frequency.

THE IMPACT OF COMPUTER AND COMMUNICATION TECHNOLOGY

Recent advances in computing have revolutionized support for control applications. The modern workstation brings a combination of advantages to the design engineer: multi-window operation allows greater individual productivity; substantial local processing power means that significant applications (for example, an optimization package) may be run with good turnaround; high-resolution screens can present the many graphical aids to controller design with very little delay. Processing power will continue to increase, and as system and database support improves the repertoire of tools available to the designer should become ever wider.

Technological advance has been as rapid in communication systems as it has been in processor power and in memory size. Data bandwidths that were unthinkable even five years ago can now be obtained using fibre optics and interface protocols such as FDDI. The cost of such communication systems is not high, usually being dominated by the cost of the earthworks required to lay ducts. The implications for control in a distributed environment are considerable. The sampling rates required to handle high frequency signal components have often ruled out control by digital processing at a central site. This state of affairs should change, always assuming that technical progress outstrips the demands made on controllers. An independent consideration is that the H_∞ design methodology should allow controllers to be developed for plants which do not admit rational models: such controllers will need to implement approximations to non-rational transformations, and digital techniques will be appropriate.

CENTRALIZED VERSUS LOCAL PROCESSING

Digital control of an application distributed across a number of sites will require a computer system at each site, if only to handle data collection and to coordinate data transmission to a central processor. The system cost of sending frequent large volumes of data for central processing is considerable: for successful control reliable synchronous communication is essential, and performance dips such as might arise from unlucky disc head positioning cannot be tolerated. It is therefore important to reduce the volume of data that is transmitted to the central site. In practice there will always be significant local processing: firstly, sampled data will require to be cleaned before transmission; secondly, digital control of mechanical systems such as valves will certainly be handled locally.

It is likely that high frequency signals requiring rapid sampling will be associated with tactical variables that control local plant, but that in order to coordinate the operation of the system as a whole some variables with lower frequency characteristics will be controlled centrally. Inevitably there will be control functions that lie at the border of the tactical with the strategic, and a decision must be made regarding their implementation. In Fig. 2 each site is assumed to have a local plant and a local computer, together with high bandwidth data communication to the coordinating central computer.

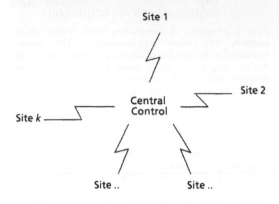

Fig. 2.

At each site the variables relevant to the local plant must be transformed. In Fig. 3 the control system for the plant at site i is represented.

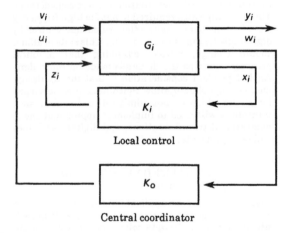

Local control

Central coordinator

Fig. 3.

Here each plant i is described by a transfer matrix $G_i(s)$, with output vector x_i controlled locally to produce feedback z_i, and w_i controlled centrally to produce feedback u_i. The controller at the central site transforms the outputs $\{w_1, w_2, ... w_k\}$ to generate the strategic feedback inputs $\{u_1, u_2, ... u_k\}$ at each site, say by implementing the transfer matrix $K_o(s)$. The control system as a whole thus comprises local tactical controllers at each site, with a central processor coordinating the overall plant behaviour. Often the choice between local and central processing for a particular variable will be clear, but there is no hard and fast line. The advantages of central control are

that greater processing power can be made available, and that data derived from other sites can be taken into account. There may be less easily quantified benefits, such as for example that additional information is instantly available to management. The drawbacks are inevitable delay, together with greater system complexity and the attendant threat to reliability.

Suppose that a control system has been designed in such a distributed environment, with plant models $G_i(s)$ and controllers $K_i(s)$ at each site, and a central coordinator $K_o(s)$. How do we estimate the robustness of such a controller, given the communication delay and computational error involved? Represent the overall plant behaviour as in Figures 4 and 5, where each system $Q_i(s)$ includes the characteristics of both plant and idealized controller. The system $Q_o(s)$ of course covers those aspects of plant behaviour that modify strategic variables, as well as the central coordinator $K_o(s)$.

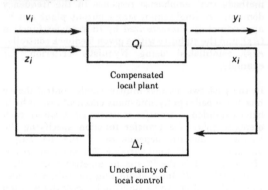

Compensated local plant

Uncertainty of local control

Fig. 4.

Wherever the behaviour of a plant or controller is uncertain, we may abstract this uncertainty from a block diagram for the compensated system. Thus if a component of plant or controller that should deliver (scalar or vector) x delivers z, we shall write $z = x + \Delta x$. In these diagrams the vector x_i consists of all signals at site i which are perturbed in such a way, and z_i the set of signals that is in fact returned to the subsystem. The transfer matrix $\Delta_i(s)$ collects such uncertainties for a given subsystem. In the same way the transfer matrix $\Delta_o(s)$ collects the uncertainties in the model of the strategic coordination system.

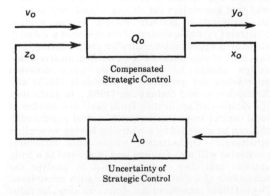

Compensated Strategic Control

Uncertainty of Strategic Control

Fig. 5.

In practice what is realised by computer will differ from the theoretical system, and for each processor it is possible to estimate bounds on the error $\Delta_i(s)$. These errors are inevitable, and arise from a variety of sources. Two that are of particular concern are errors in calculation and errors due to delay. The former can often be reduced if more processing power is made available: they may be estimated by numerical analysis. The latter reflect the extent by which a discrete model with given time step fails to approximate a continuous system, and will usually increase with frequency.

The overall system can now be realised by a transfer matrix

$$Q(s) \quad = \quad \begin{bmatrix} Q_{11}(s) & Q_{12}(s) \\ Q_{21}(s) & Q_{22}(s) \end{bmatrix}$$

where signals in the ideal model $x \equiv \{x_0, x_1, x_2, .. x_k\}$ and propagated signals $z \equiv \{z_0, z_1, z_2, .. z_k\}$ are related by $x = Q_{22}(s) z$.

The transfer matrix $Q(s)$ models the system so as to take account of the transformations carried out by computer. The component $Q_{22}(s)$ represents the transfer function by which computer generated error is fed back through the nominal compensated plant model. In any realistic case this matrix will not conform to the structural constraints on the uncertainties, for example errors in handling signals at the central controller will be propagated to local feedback loops. There is therefore no simple formula that specifies the structured singular value. In such a case the robustness of stability and performance of the system as a whole under computer errors must be gauged by estimating $\| Q_{22}(s) \|_\mu$.

There will of course be errors in the model $Q(s)$ of the system other than the structured uncertainty considered above. The purpose of the above analysis is to treat the specific errors associated with computer transformation. Provided that compensated plant stability is robust under such errors overall system stability will follow. It is shown in Doyle, Wall and Stein (1982) that robustness of performance may also be established by considering $\| Q_{22}(s) \|_\mu$.

RESPONDING TO THE UNEXPECTED

In the previous section we considered how to estimate the effect of uncertainties in computer processing on the stability of a distributed control system. The approach lay within linear control theory, and if the model is to be valid tight operational constraints must be satisfied. The model assumes synchronous interaction between plant and controller, with the latter comprising computational as well as mechanical components. The demands on the computer system are severe: robust communication protocols must guarantee to deliver data on schedule; algorithms that transform control variables must be susceptible to accurate error analysis. On the other hand, provided that operational constraints are maintained, the demands on the system are predictable.

In practice mechanical or computational component failure will threaten the operational constraints, and the system will be monitored to identify potential breaches. In a complex environment there will be many exception conditions, each of which will require specific incident analysis routines to be scheduled. The requirements for handling such incidents include the following:

a) the wide range of potential incidents will require that routines are scheduled from disc;

b) the structuring of such incident analysis software must be coordinated with the design phase;

c) incident analysis will involve access both to the design database and to data collected in real-time by the control system;

d) regardless of where an incident is detected, the analysis routine must interact with the distributed environment.

Note that incident handling will interact both with the design database and with the real-time control system. Further, the need for distributed analysis and response will generate communications traffic. It is important that the real-time operating system provides a model that allows the designer to plan with incident analysis in mind. Only then can it be guaranteed that control system performance is as far as possible robust when such an incident takes place.

CONSERVATIVE VERSUS STATE-OF-THE-ART DATABASE TECHNOLOGY

To a computer scientist with a background in conventional database applications the world of distributed real-time control is intimidating. Engineering a database management system is a major project which should not be undertaken lightly. Commercially available DBMS are the result of many man-years of effort, and their adoption brings significant advantages: the better products are reliable and well-documented, and have been tuned to handle gigabytes or even terabytes of data with acceptable performance. Although there are still large business databases that are managed using CODASYL systems, the relational model dominates recently developed DBMS. There is now an SQL standard for data definition and data manipulation, and many commercial systems support schema modification as a matter of routine.

It is interesting to reflect that fifteen or so years ago the claim was made that relational DBMS could not be engineered efficiently. The change has come about through continuing research in many areas. Two of the best known systems available today are INGRES and IBM's DB2, each the direct offspring of a research project of the 1970's: fundamental papers on many different topics arose from these two projects, see for example Stonebraker and Wong (1974), Selinger and others (1979), and Gray and others (1981). It is probably realistic to look for a ten-year development period from the first practical research systems to a viable software product.

What requirements should a DBMS meet if it is to be suitable for use in a control environment? In particular, the DBMS must support both the design and the operation of a control system in such a way that interaction between the two is natural. The following considerations are taken from earlier sections, and the list is far from exhaustive:

a) it must be possible to define and manipulate objects with complex structure, and structural change itself must be easy;

b) during the design phase there must be support for version maintenance, together with logging and recovery of long transactions;

c) the scheduler must give good response to transactions initiated by unforeseen events without prejudicing deadlines for the synchronous processing of large volumes of data;

d) the database must be integrated with a distributed real-time operating system, ideally one with a predictive performance model;

e) if software products are to be developed then standards will be needed for languages and data models.

The relational data model is generally agreed to be inadequate for CAD applications, and much has been written about alternatives. Within the general framework of the ANSI/SPARC architecture the functional data model of Shipman (1981) offers richer semantics, and some attention has been paid to it. Even more interest has been aroused by Database Programming Languages (Atkinson and Buneman, 1987) and Object-Orientated Database Systems (Dittrich, 1986). There is a substantial overlap between these active research areas, and prototype systems now being developed share many common goals.

It is easy to find recent research publications that relate closely to the list of requirements for a DBMS suitable in control applications:

a) support for objects with complex structure, and for versions of the schema describing such objects (Kim and Chou, 1988);

b) design of a recovery manager for nested transactions (Korth, Kim and Bancilhon, 1988), (Rothermel and Mohan, 1989);

c) high-performance object-orientated database systems (Maier, 1989);

d) a real-time operating system kernel with performance prediction (Tokuda and Mercer, 1989), scheduling database transactions subject to deadlines (Abbott and Garcia-Molina, 1989), (Carey, Jauhari and Livny, 1989);

e) a (belated?) search for standards for object-orientated databases (Atkinson and others, 1989).

The state of research on such problems is at roughly the stage that relational database technology had reached in the late 1970's. It is likely that it will be ten years or so before this research leads to integrated systems that can support distributed real-time control. Although real-time control systems provide a major challenge to DBMS designers, the requirements are shared with many other applications. For example, video service in an office information system involves synchronous delivery of large volumes of data, though the penalty for missing deadlines may be less severe than in the case of the real-time control of a nuclear power station. The time is probably not yet ripe for the development of (as opposed to research on) DBMS explicitly for real-time control. On the other hand, database researchers are always on the lookout for databases that exhibit properties relevant to their current interests - the pilot application of System-R to a number of databases in the United States in the late 1970's upset a number of the assumptions on which

software development had been based. A fruitful outcome of this Workshop would be for research workers in databases and in distributed systems to continue to meet with control engineers, and ideally to cooperate in investigating research problems within a realistic context.

CONCLUSION

Advances in technology are revolutionizing real-time digital control. In particular, the rapid increase in readily available communication bandwidth makes it feasible to shift large volumes of data. In a distributed control environment there will be arguments for both local and central computer transformation of control variables: in either case there will be uncertainties in implementing such transformations. Provided that the performance characteristics of hardware and software can be modelled so as to quantify such uncertainties, their effect on the stability and performance of the control system can be gauged using structured singular values. Analysis of this kind during design should help to decide how to distribute control.

ACKNOWLEDGEMENTS

I am particularly grateful to Jan Maciejowski, who lent me copies of Francis (1987) and Maciejowski (1989) to provide a crash course in Control Theory. He also spent a substantial time in resolving my difficulties. My continuing ignorance is no fault of his. Keith Carne helped me to get to grips with Hardy spaces, and pointed me at Garnett (1981) and Partington (1988). Jean Bacon and Raphael Yahalom read drafts of this paper and made many helpful comments.

REFERENCES

Abbott, R., and H. Garcia-Molina (1989). Scheduling Real-Time Transactions with Disk Resident Data. *Proceedings of the 15th International Conference on Very Large Data Bases.*

Atkinson, M.P., and O.P. Buneman (1987). Types and Persistence in Database Programming Languages. *ACM Computing Surveys*, **19, 2.**

Atkinson, M.P., F. Bancilhon, D. DeWitt, K.R. Dittrich, D. Maier and S.B. Zdonik (1989). The Object-Oriented Database System Manifesto. *Proceedings of DOOD 89, Kyoto, Japan.*

Bernstein, Philip A., V. Hadzilacos, and Nathan Goodman (1987). *Concurrency Control and Recovery in Database Systems.* Addison-Wesley.

Carey, M.J., R. Jauhari, and M. Livny (1989). Priority in DBMS Resource Scheduling. *Proceedings of the 15th International Conference on Very Large Data Bases.*

Dittrich, K.R. (1986). Object-oriented database systems: the notions and the issues. *Proceedings of the International Workshop on Object-Oriented Database Systems, IEEE CS, Pacific Grove, California.*

Doyle, J.C. (1982). Analysis of feedback systems with structured uncertainties. *Proceedings of IEE, Part D,* **129,** 45-56.

Doyle, J.C., J.E. Wall, and G. Stein (1982). Performance and robustness analysis for structured uncertainty. *Proceedings of the IEEE Conference on Decision and Control, Orlando, Florida*, 629-636.

Edmunds, J.M. (1979). Control system design and analysis using closed-loop Nyquist and Bode arrays. *International Journal of Control*, **30**, 773-802.

Francis, B.A. (1987). *A Course in H_∞ Control Theory*. Springer-Verlag.

Garcia, Carlos E., David M. Prett, and Manfred Morari (1989). Model Predictive Control: Theory and Practice - a Survey. *Automatica*, **25**, 335-348.

Garnett, J.B. (1981). *Bounded Analytic Functions*. Academic Press.

Gray, J. and others (1981). TheRecovery Manager of the System R Database Manager. *ACM Computing Surveys*, **13, 2**.

Kim, W., and H. Chou (1988). Versions of Schema for Object-oriented databases. *Proceedings of the 14th International Conference on Very Large Data Bases*.

Korth, H.F., W. Kim, and F. Bancilhon (1988). On Long-Duration CAD Transactions. *Information Sciences*, **46**, 73-107.

Kreisselmeier, G., and R. Steinhauser (1983). Application of vector performance optimization to a robust control loop design for a fighter aircraft. *International Journal of Control*, **37**, 251-284.

Lampson, B. (1986). Designing a Global Name Service. *Proceedings of the Fifth Annual ACM Symposium on Principles of Distributed Computing*, pp. 1-10.

Lindsay, B.G. and others (1979). *Notes on Distributed Databases*. IBM Research Report RJ2571.

Maciejowski, J.M. (1989). *Multivariable Feedback Design*. Addison-Wesley.

Maier, D. (1989). Making Database Systems fast enough for CAD Applications: *Object-Oriented Concepts, Databases, and Applications*. ACM Press.

Partington, J.R. (1988). *An Introduction to Hankel Operators*. LMS Student Texts 13, Cambridge University Press.

Rothermel, K., and C. Mohan (1989). ARIES/NT: a Recovery Method based on Write-ahead Logging for Nested Transactions. *Proceedings of the 15th International Conference on Very Large Data Bases*.

Selinger, P. and others (1979). Access Path Selection in a Relational DBMS. *Proceedings of the 1979 ACM SIGMOD International Conference on Management of Data*.

Shipman, D.W. (1981). The Functional Data Model and the Data Language DAPLEX. *ACM Transactions on Database Systems*, **6, 1**.

Stonebraker, M., and E. Wong (1974). Access Control in a Relational DBMS by Query Modification. *Proceedings of the ACM Annual Conference*.

Tokuda, H., and C.W. Mercer (1989). ARTS: a Distributed Real-Time Kernel. *ACM SIGOPS Operating Systems Review*, **23, 3**, 29-53.

Youla, D.C., H.A. Jabr, and J.J. Bongiorno (1976). Modern Wiener-Hopf design of optimal controllers: the multivariable case. *IEEE Trans. Auto. Contr.*, **AC-21**, 319-338.

PROCESS CONTROL:
HOW TO PROCESS DISTRIBUTED DATA
IN LESS THAN A SECOND?

S. Sédillot

INRIA, Domaine de Voluceau, Rocquencourt, BP 105, 78153 Le Chesnay, France

Abstract. Physical and natural distribution is also a property for large process control realisations. The next step in automatization concerns the optimized designs of data flows and data processing on the floor. The most performant architecture with regards to database distribution, is certainly to distribute data, that is, to consider the usage of cooperating servers. ISO is actually working on a communication support to distributed transactions that must be atomic. The Distributed Transaction Processing protocol is an open protocol in that it is a support to the transaction concept but is not meant to interfer whatsoever in the containt of the transaction. Between the times a transaction starts and terminates, an application process may use indifferently RDA services, MMS services or file transfer services. Presently, the so called application layer may be considered as a pool of protocols that offer specific services to the application process. However, there remains a work to be accomplished in ISO under the pression of network users with regards to interoprability between protocols. This is a goal that can be achieved in the determination of the standardized so called protocol "profiles" options. To this aim, all potential users should participate in this work at SPAG, in ESPRIT projects within the European Community and in ISO working groups.

INTRODUCTION

The real-time world is by excellence a word of distribution : Robots are distributed in the cell, manufacturing cells are distributed throughout a shop and shops are physically distributed throughout the factory. Physical and natural distribution is also a property for large process control realisations. However, distances generally do not exceed few kilometers. This is a fact. Factories or process control automatization has been made possible because, since very few years, local area networks, field busses, token rings, token bus, ethernet and FDDI networks technologies are available through reliable and relatively cheap products and enable the physical interconnection of workstations, cells controllers and mainframes. The next step in automatization concerns the optimized designs of data flows and data processing on the floor in order to improve both reliability and response times. More reliability is an obvious mean to improve benefits since it reduces the cost of machine unavailibility. Improvment of response time is the only way towards more schedule flexibility.

On the shelf local area network physical access methods and data transfer standards permit only to transmit data between two computers, but none of them offers services specifically suited to the transmitted informations semantics. This is a good point because, being general purpose tools, they may be used by higher level standards that are meant to be the interface between application programs and the networks.

A standard, in the network world, is a protocol, that is, a set of rules that govern the exchange of informations.

There is now a large effort in the definition of standards that are aimed at providing secure specific high level communication services to support database remote access, task remote monitoring distributed transactions processing and remote file exchanges. However, some issues are still open with regards to the interoperability of these standards. This paper is an attempt to analyze to what extent interoperability of application standards is desirable for processing data in the factory, to give an overview of these application standard specificities and to emphasize major requirements in such interoperability.

The first section analyzes the data flows related to scheduling, reporting and monitoring in the plant floor. The plant workstation data processing functions are taken as a template to enlighten the requirements on communications tools. Particular attention is paid to the implementation of the protocols and application software in terms of task and message scheduling requirements.

The second section shows the requirements in remote monitoring in order to obtain flexibility in the factory. More precisely remote loading of programs and tasks remote execution monitoring are adressed and it is shown that the Message Manufacturing Specification Protocol fulfills the requirements with the help of the Association Control protocol.

The third section gives an overview of the ISO reference application layer architecture in order to familiarize the real-time engineer and give him a hint of the possibilities it can expect from the communication protocols that are or will be offered to him in the next five years.

The fourth section adresses database access. A clear distinction is made between remote accesses to a centralized database and cooperation among distributed Databases. It is shown that the Remote Data Access protocol is aimed to provide a support for communication related to the first approach but must be used together with the Transaction Processing protocol if the second approach is taken.

The conclusion emphasizes the benefits of using the TP protocol with regards to redundancy of data, multiple copies consistency and recovery upon failures.

Finally, special attention is given to the criteria that a real-time user may demand from the vendors, with regards to communication protocols interoperability.

DISTRIBUTED DATA PROCESSING REQUIREMENTS FOR CONTROL AND MONITORING APPLICATIONS

A typical application for a workstation is sensing and inspecting its subsystem, i.e the group of equipment controllers for which it is responsible. It implies large storage buffer, control loops with response times ranging from a second to few seconds. The trend is that workstations will reach acceptable speed and memory to trigger equipment controller operations. However there are two domains in which workstation shall never be autonomous : They are , on one hand, the middle and long term scheduling functions, and, on the other hand, the general floor reporting of statistics on pieces flow, on quality control, on human interventions and on floor environment such as temperature, electric power, etc.. Such information are sampled, occasionnally preprocessed for local operations like feedback on local actuators, but also reported to the cell controller. Likewise, informations related to scheduling are received fom the cell, or, even directly from the shop controller. Thus, it appears obvious that networking activity is a must in the workstation toolkit. The same reasonning holds for a cell and a shop controller although response times in data processing may usually increase.

Precisely, as far as networking is concerned, a control application program shall be ready to receive data or files, store them, possibly process them so as to extract either a specific view of part of the process (reporting) or a set of orders (scheduling) and transfer it to the ad hoc controller, be it at a higher or a lower level (BAH, 87) in the factory hierarchical model (Figure 1).

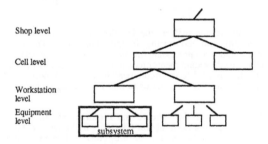

Shop level

Cell level

Workstation level

Equipment level

subsystem

Fig. 1.   Real-time environment model.

Note that one should not think of reporting only as a strictly hierarchical function based on polling and inducing periodic trafics : In our example, the cell controller could poll each equipment controller of its subsystem every minute. Reporting can also occur due to an abnormal situation detection in any level of the factory. In such cases, periodic polling by the superior is out of sense. One shall consider such a reporting as an event that suffers no delay in processing and in possible transmission to the level above.

In conclusion, it is obvious that a computer at any level must be able to process, receive and transmit data with at least two different priorities to its superior(s) as well as to its subordinate(s) controllers in the hierarchy. Depending on this priority, response time for transmission or processing ranges from the order of the second to few tens of seconds. Note that the nature of the data flow and the priority are related : Alarm data flows consist of short data and their occurence cannot be predictable in time. Conversely, statistic reporting generates large data flows whose occurences may be predictable in time. Coexistence of different ranges of response times implies a priority attribute in messages transmitted over the network(s) and corresponding scheduling of messages. There are two types of software design related to the application tasks and to the tasks that implement communication protocols that are meant to meet message scheduling : One can design as many communication processes (they are called "tasks" in the litterature related to real-time executives) as there are potential message priority types (Figure 2).

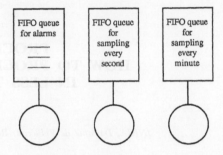

Fig. 2.   Priorities assigned to processes.

An other possibility is to design a single task (also called a "server") that processes ordered messages from a queue. Messages are ordered in the queue according to their priority (Figure 3).

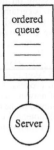

ordered queue

Server

Fig. 3.   Priority assigned to messages.

The later solution has the advantage to avoid task context switching in the computer, which is pure overhead, eventhough computers based on RISK architectures minimize such overhead. However, this choice might not be possible because transmission requirements may differ significantly depending on the data structures that are transmitted. For example, a bulk transfer (5 Moctets) require segmentation and flowcontrol whereas a short data unit such as an alarm is a ponctual transfer that requires only error control. Thus, the design of processes and queues is a trade-off between the minimization of the process number and the number of data transfer protocols.

One must be aware that the standard protocol class IV (ISO, 84) offers a simple transmission service (T-DATA) to the application. The transport protocol however segments too long messages, checks the sequencing of segments upon reception, reassembles the segments into one message, controls the flow of segments, checks if any transfer error occurs and eventually repeats the transmission which induces an overhead in the order of 3000 instructions per segment. Eventhough the overhead is smaller for a short message (a single segment), the numbering of segments and the checking for authorized transmission induces a non neglectable overhead. On this ground, the MAP (MAP, 87) approach for an enhanced architecture, the suppression of the transport protocol and the use of an acknowledged Datagram protocol is reasonnable.

MONITORING AND FLEXIBILITY

Monitoring should be taken in a broad sense : that is, it should imply the downloading of workstations and controllers with programs and parameters, as well as the dynamic change of parameters. As can be well observed in a plant floor, software is installed each day, sometimes each hours, even more frequently, according to plant, shop and cell schedules. It is only when such operations such as are downloading and dynamic changes of parameters are easily managed and rapid that one can consider a plant floor as a flexible one. Note that we insist on the fact that we are not adressing here long term schedules that take a long time to be computed but rather very short term schedule changes. Such changes do not necessarily originates from the top level of the factory but also may be subsequent to failure of pieces, defaults in parts etc... that is subsequent to some event detected through control at lower levels. It is then obvious that software changes should be performed very rapidly, so as to avoid to let machines do erroneous or useless and costly work. Changes of configuration in terms of software imply :

- Downloading of programs and data set (parameters),
- Start of new tasks,
- Stop of old tasks,
- Checking through status reporting that new tasks are in the active state.

This is exactly the type of services that are offered by the Message Manufacturing Specification (MMS, 87).

Moreover, it may happen that two or more cell controllers are narrowly dependant on one another. For example, in an assembly plant, functions and timings performed by a cell controller downtream the pieces conveyor depend on functions and timings performed by the cell controller upstream the pieces conveyor. The consequence of this is that changes in the set of the active tasks in one cell controller implies synchronized changes (with a delay that depend on the pieces conveyor speed) in the other cell controllers. There are two ways of synchronization : a first one is to prepare (download a new set of tasks in both cell controllers and order local systems that they should switch from their current set of tasks into the new set at a given time. Note that switching times are different for each cell controller and that it implies physical local clock synchronization. The second way is to download the new sets of tasks and use message passing between the cell controllers whereby one cell controller orders the other cell controller to switch -possibly after a delay- from the old set of tasks into the new set. The reception of the message by the second cell controller is to be considered as an "event" in terms of execution. MMS provides the tools to do so : it offers services that

- Create an event description in a remote machine,
- Attach a task to an event descriptor and forces one or more tasks to wait until the event switches from one state into another,
- Forces a remote event state transition.

MMS does not provide support for local clocks synchronization. This is left to the user care.

Concerning the downloading, MMS offers services that enable one computer to order to an other computer to trigger the loading into its memory or on any of its memory support of a domain localized on any computer . A domain is a set of structured blocks of memory. A domain may contain programs or/and data.

MMS protocol utilizes the presentation basic (kernel) transfer facility on an association established between two distributed application processes.

An application process may establish and release an association with a peer application process through usage of the Association Control Service Element services (ACSE, 88). An association is a set of resources : buffer, queues, variables that are used by protocols to transfer messages.

THE ISO-OSI APPLICATION LAYER

It appears from above that an application process shall call upon different kinds of services in order to perform its functions : In the example above, the application process shall use specific services for establishment of an association with a peer application process and specific services for monitoring and control. Before adressing communication support for remote database accesses and transactionnal operations, it is necessary to point out architectural aspects of the communications standards that may impact closely performances in terms of response time.

The International Standard Organization (ISO) Reference Model for Open Systems Interconnexion (Figure 4) is based on peer-to-peer communication and a layered architecture. There are six layers. An open system is a system that contains the minimum number of standard communication protocols to transfer and receive data.

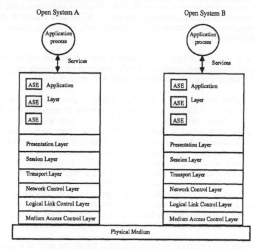

Fig. 4.   The ISO OSI Reference Model for peer-to-peer Open System Interconnections.

Each layer offers added value to the layer below : medium access, application-to-application reliable transfer (they are multiplexed on computer-to-computer transfer), synchronisation of exchanges (half duplex, full duplex, checkpointing), encoding rules.

The seventh layer, called the application layer, offers services directly to the application process.

The application layer is the last to be defined and is currently subject of a great amount of work in terms of architectural design (ALS, 89).

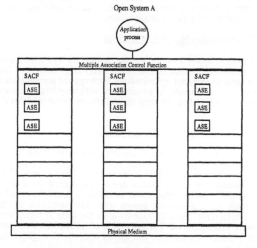

Fig. 5.   Application Layer Structure.

The problems to solve are two-folds : an application process may, in the course of its lifetime, use quite different sets of services like Message Manufacturing services, file transfer services, Remote Data Access services and transactionnal services. One normally expects that these different sets of services are provided by different protocol machines, each of them having different encoding of messages, different rules governing message sequences, different state informations that are used for checking the sequences of messages and for recovery upon failures. Consequently, each set of service and its protocol are grouped in a so-called Application Service Element (ASE). Thus, application process can interleave requests on different ASEs ; The ideal would be that the application process establishes is an association with optionnal quality of services that match all the kinds of exchanges that it expects to perform through the use of ASEs and that meet its requirements in terms of synchronisation, reliability, security and response time.

Note that the Single Association Control function within the application layer controls that the interleaved requests respect the sequencing rules related to each ASE requirements.

Another point is that a single Application process may find it necessary to carry on reliable exchanges of messages with more than one application process : For example, an application process may have to gather from two or more workstations statistics that have relationships one with the other. Another example is the case where the application process need to broadcast orders. To fulfill such requirements, the application layer structure supports a multiple association control. The Multiple Association Control Function within the application layer coordinates the sequencing and consistency of the application process requests on all the associations for which it is responsible. Note that a different set of ASEs may be used by the application process on each associations.

Given this structure in the application layer, the goal of a good design is to find out compatible options in the use of the ASEs and in establishment of several associations so that performance requirements are met. This is not trivial because ASEs have different requirements on the association options. As an example, suppose an application process wishes to use three ASEs : if options regarding the association (half duplex, full duplex, checkpointing) are compatible, the application process may establish a single association and interleave requests to the three ASEs. Conversely if options regarding the association are not compatible, the application process must establish two or three associations. This is costly in resources.

REAL-TIME DATABASE ACCESSES

There are still no in situ architectures where a workstation accesses a database with real-time constraints. This deserves some in depth consideration and one should clearly argue what for real-time databases might be used.

Centralized database and local access

A staight forward example is the computation of a long term manufacturing schedule that needs a very large amount of data in order to carry out optimizations. Such type of computation is generally performed on a mainframe with very large memory. Subsequent orders for shop controllers are extracted from the result and transmitted.

Centralized database and remote access

Up to now, I have heard of two situations in which the factory or the control process technical staff was thinking of Databases as the thing they need. It is necessary to keep a trace of the conditions in which pieces are manufactured in order to establish quality reports and in order to fetch the possible origin of defectuous pieces that are returned to the manufacturer, possibly months after the time at which they were manufactured. The collection of informations on parts and pieces is enormous and grows with the throughput and with time.

The advantage of using databases instead of files to save the collected informations is mainly in terms of response time. As a matter of fact, data manipulation languages offer the user with the possibility to do selections of informations that respond to several criteria at a time and therely minimize the research delay. Moreover, the physical distribution of data may be defined so that data related to the same subject are localized in a same or in few tuples which minimize the access time to data. An other point of view was expressed to me, namely that

a database might help in the way to configurate workstations and controllers. It is felt that configuration parameters may become so numerous that a single operator cannot manage to take in charge their processing for each workstation and controller but rather, each workstation and controller should read in a database the parameters that it needed.

In both example, in terms of communication, we have "clients" (workstations and controllers) that send or request informations to a "server" where the server owns the database.

ISO is currently on work in this area : the RDA protocol (RDA, 89) is meant to provide access to a remote centralized Database (a server). The goal of this protocol is essentially two-folds : First, provide a client user with services for establishing that is a set of resources on the client computer and on the Database server a dialogue, and select the database it wishes to access. Note that a Database server may offer access to one or more databases. This allow a workstation to access sequentially several database, each of which may be the property of different staffs in the factory. This is important because although different databases may contain replicated data, access authorization (specially for writing and modifying) may be different.

The second objectif of RDA, as it stands, its to classify remote data accesses into four services : Execute a statement, Define a statement, Invoke a previously defined statement, Erase reference to a previously defined statement (Drop). A result or an error message is returned for each service.

The Database language SQL specialization in RDA is also under work at ISO : SQL Data manipulation statements encoding is under standardisation in the ASN1 language. An encoded SQL statement may be conveyed to the server through the Execute service parameters.

In terms of performance, the actual definition of RDA has some severe drawbacks : a client of a multiple database server should be allowed to access simultaneously several databases. A more convivial language than SQL may be necessary for human operator that are not specialists of databases. RDA services might be not adapted to an object oriented language for example. Intermediate results (one per statement) might be pure overhead when a client is be interested only in the result of a sequences of statements. Lastly, RDA does not offer services for Data Definition statements. However, this is obviously one function that is necessary throughout the factory life time.

The access time to a centralized database might be prohibitiv, at least for configuration. One must be aware that the response time depends on the number of computers and network segments over which access requests and related responses have to flow. Moreover, this flow induces overhead in passed through computers that will have their performance decreased.

Distributed Databases and remote access

The most performant architecture with regards to database distribution, is certainly to distribute data, that is, to consider the usage of cooperating servers. The arguments in favor of data distribution are the followings.

Regarding the above example, one might observe that decisions are not always taken in a single place on a plant flour : a human being in charge of a shop or even of a cell may also take decisions. Moreover these decisions do not always rely on orders from the above level but also and, sometimes, only, on reportings from the lower levels (equipments and workstations). Such reporting may need to be totally or only partially reported on the layer above for further study. Then the natural approach is to collect reportings and save them in a database located where the decision has to be made and have several processes that each extract specified data and reports them either to the operator, or to other the levels for further controls. At this condition, efficient automatization may be obtained.

Then one has to face such operations as copying of a data set into several databases, split local data into remote databases in a reliable fashion, that is, with the guarantee that local data are not forgotten (erased) before they are secured wherever they must be transmitted.

Another domain in which reliability should be the main goal is the multicast or broadcast of operational data. For example, co-related orders for changing parts in different equipments may be issued by a cell controller to one or more workstation. Either all of none of the changes shall be performed. The cooperation of two or more distributed tasks that must be either all performed or all cancelled is called an atomic transaction. This concept of the transaction in the factory has the same requirements that in the banking area if one notes that orders to remote workstations are equivalent to simultaneous orders to Wall Street and banks for the same account : they are in both cases, data manipulations because, in the case of the factory orders to equipments are created and transmitted to the end equipment controller as pure data. Note also that beside the property of atomicity, execution of such transactions may require a specific property which is called isolation. That means that during transaction A execution, no other transaction should be authorized to read or modify part of the data accessed by A.

ISO is actually working on a communication support to distributed transactions that must be atomic. Essentially, the Distributed Transaction Processing protocol (TP, 89) offers an atomic termination of a transaction span over a tree made of two or more associations (Figure 6) .

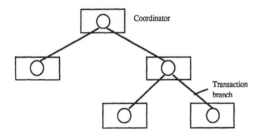
Coordinator

Transaction branch

Fig. 6.   Transaction tree span on five computers.

The underlying mechanism to achieve the termination phase of a distributed transaction is the well known two-phase commit protocol. It is made up of four waves of exchanges, and is controlled by one of the partner computers, called the commitment coordinator. The four waves are defined as follows :

a) The commitment coordinator initiates the termination phase by issuing to its partners a "prepare" request.

b) Each of the recipient Open System shall reply to the "prepare" request by either accepting or refusing to proceed. Accepting means that the Open System has ensured that, later on, it will be possible to secure bound data (bound data are data modified by the transaction). If this is possible, the Open System notifies the commitment coordinator that it is "ready". The Open System is also said to have put its bound data into the "ready" state : from this state, the bound data may be released in either the final or the initial state, according to the commitment coordinator's decision (see next).

If getting "ready" is not possible, the Open System rolls back the distributed transaction.

c) Whenever the commitment coordinator has been notified by all its partners that they are "ready", it orders them to "commit", i.e. to release their bound data into their final state.

d) After being ordered to "commit", each Open System secures its bound data into their final state and notifies the commitment coordinator of the completion of its own part.

Consistent termination (either rollback or commitment) on all computers is guaranteed in spite of possible communication failures or computer crash by means of logs of the termination process current state on secure storage by all computers. Therefore the TP protocol implies that computers do have a secure storage access. If a communication failure or a node crash occurs during the termination process, an Open System recovers his secured log and consequently polls the commitment coordinator in order to obtain the final outcome of the transaction. It is clear that this polling may take a certain time which includes a discrepancy in the delivery of co-related orders or data to the involved application processes. For this reason, a correct design of real time network architectures implies deterministic transmission times and, if strategic, redundancy of processors and network segments Because local area networks exhibit a high reliability and high speed (Transfer time of 1 millisecond is reachable), this discrepancy should be very rare.

One outstanding potentiality of the TP protocol is that it allows management of consistent replications of data : Multiple copies on different computers can be kept continuously identical provided that they are updated within a single transaction. Given this, one should compare the cost of securing a mainframe to the cost of running statandard softwares in duplicated small computers.

The Distributed Transaction Processing protocol is an open protocol in that it is a support to the transaction concept but is not meant to interfer whatsoever in the containt of the transaction. Between the times a transaction starts and terminates, an application process may use indifferently RDA services, MMS services or file transfer services.

Moreover, TP has its own association management, maintains an association pool and can affect an existing or a new association to the support of a transaction branch between two computers.

CONCLUSION

In conclusion, an application task must select the appropriate protocols to support its operations. Presently, besides the protocols related to the five lower layers that are concerned by the quality of transport on one association between two computers and the sixth layer that is concerned with the encoding definitions, the so called application layer may be considered as a pool of protocols that offer specific services to the application process.

However, there remains a work to be accomplished in ISO under the pression of network users with regards to interoprability between protocols. Protocols at the application layer are already compatible with respects to the services that they offer since they are related to quite different functionalities. But these protocols are born somewhat independently from one other, thus each of them is defined as being first invoked with a service that establish an association. Clearly, this is not optimized since it compells an application process to use several associations (partially redundant resources) to communicate with a peer application process. It should be possible to interleave, on the same association, exchanges of data related to a large set of protocols.

Moreover, some protocols have some restrictions on the flow of data exchanges, some do not and leave it to the application process. As a matter of fact, TP offers two modes of data exchanges : full duplex or constrained by the ownership of a token. MMS allows asynchronous confirmed requests. RDA allows requests with result or error response only at the client side of an association. None of these modes are incompatible but it is clear that token passing to get the right to transmit might result in useless overhead in processing.

In all cases, application protocols should require the minimum flow constraints from the lower layer so that the same lower layer negotiable service options (half duplex, full duplex, token management and checkpoints management) constitute a set that may support a maximum number of application protocols and let the user master its own way to trigger exchanges. This is a goal that can be achieved in the determination of the standardized so called protocol "profils" options. To this aim, all potential users should participate in this work at SPAG, in ESPRIT projects within the European Community and in ISO working groups.

Moreover, databases vendors that presently prepare next releases with the transaction and the commitment concept must take into account the Remote Data Access and the Transaction Processing services so that users might be provided with networking facilities.

REFERENCES

[ACSE, 88] ISO 8649 IPS/OSI. Service definition for the Association Control Service Element.

[ALS, 89] DP 9545. Revised text : 2-6, 1989. Information Processing Systems. Open Systems Interconnection. Application Layer Structure. ISO/IEC/JTC1/SC21.

[BAH, 87] Connectivity in the Factory. J. Bahr, Unix Review, June 87.

[ISO, 84] DIS 8072. Information Processing Systems. Open System Interconnecion. Transport Service Definition.

[MAP, 87] MAP 3.0. Specification, May 1987. European MAP. Users Group editors and Process Control Architecture, Draft D, June 87. AFNOR. Editor.

[MMS, 87] Manufacturing Message Specification. Service Definition, ISO/DIS/9506/1. Dec. 87.

[RDA, 89] Interim text of 2nd DP 9579. Generic Remote Database Access Service and Protocol.

[TP, 89] DP 10026. IPS - OSI. Distributed Transaction Processing.

SPECIFICATION AND IMPLEMENTATION
OF CONCURRENCY IN A
SIMPLE DATABASE SYSTEM

B. Gliss and J. Gorski*

Max-Planck-Institute, 7000 Stuttgart 80, FRG

*The work carried out while on leave from Institute of Informatics,
Technical University of Gdansk, Gdansk, Poland*

ABSTRACT The implementation of a simple database system serves
as an example for a concurrency control protocol and its
implementation within the VMS operating system environment. The
protocol specification in terms of communicating finite state
machines and the implementation of the underlying interprocess
communication mechanism are presented.

KEYWORDS database; concurrency; protocol; interprocess
communication

INTRODUCTION

The implementation of a simple database
system PDBS [Gliss 1988] serves as an
example for a concurrency control protocol
and its implementation within the VMS
operating system environment. The PDBS
system is characterized by the following
properties:

- support of the relational data model,
- availability of some nonstandard
  extensions, mainly in the area of
  available datatypes and the possibility
  of using scripts for database
  manipulations,
- a query by example like human interface
  with strong user control of final
  resuts through browsing and editing
  facilities.

The system has been developed applying the
outside-in approach with strong emphasis
on user iterfaces. The users have been
involved in the development very early and
kept interested throughout the whole
process. It was achieved by forming a user
group equipped with some power of
decision, providing early prototypes,
involving the users in system testing, and
most important, considering the users'
feedback seriously. The system design is
based on the object oriented approach
[Booch 1986], which means that the system
is composed of objects. The objects
interface to each other by invoking
operations. In our environment, based on
VAX-Pascal [VMS Pascal 1987], we had two
basic patterns for object implementation:

(1) an object is a collection of
procedures grouped around a static data
structure which represents the object's
internal state. The MODULE..END, INHERIT
and ENVIRONMENT extensions of VAX-Pascal
provided for partitioning into
specification and implementation parts,
visibility restriction and separate
compilation of such representation;

(2) an object as a process (task) and
its operations are implemented by message
passing (or any related mechanism). In our
environment this had to be done by
explicit use of the VMS operating system
services (Pascal is not a concurrent
language). The most straightforward
implementation of the operations is to
simulate each invocation by two messages:
the call message and the return message.
However, for sake of efficiency this can
be then refined and some messages (those
which are meaningless and carry no
information) can be discarded. The result
is a protocol which is run by the group of
processes (objects).

In the rest of this paper we present the
protocol for those objects of the PDBS
system which are implemented as processes.
We also present the IPC module which
implements the message passing mechanism
based on the VMS system services. More
information about software technology in
the development of the PDBS system is
presented in [Gliss and Gorski 1989].

THE PROCESS STRUCTURE

The system was designed with corse grained
concurrency in mind. The reason for this
is a compromise between a better use of
system resources through concurrent
processes and the cost incurred when
creating separate tasks; the base
operating system [VMS Pascal 1987] does
not support lightweight processes. The
system structure involves the following
top level objects that run under the

control of a monitor process PDB spawned
by the VMS's user process:

- the user interface UIF. It represents
  the facilities and the interface
  offered to the system users hiding the
  screen layout, textual representation
  of messages to be displayed, etc. It
  also manages this part of the system
  state that is responsible for the
  output of results;

- the database definer DDF. It hides the
  structures necessary to define and
  manipulate relational schemata. It
  maintaines a dialouge with the user via
  UIF and stores the current database
  schemata;

- the database manipulator CPS. It
  organizes the dialouge with the user
  for inserting or retrieving data items
  in the database and hides the knowledge
  about how this dialouge is organized.
  It uses the services of UIF and stores
  the results of the user dialouge as a
  transaction expressed in a suitable
  intermediate language;

- the database manager DBM. It
  manipulates the database for the user
  interacting with CPS. It interprets the
  intermediate language program produced
  by CPS and handles the user's data
  within the context of that transaction;

- the utility manager UTL. It runs
  various utilities (e.g. a sort utility)
  for the user.

The lifetimes of DDF, CPS and UTL do not
overlap, i.e. the user either defines
(modifies) the present database schema
(working with DDF) or manipulates the data
within the present schema (trough CPS), or
runs some utilities (UTL) against the data
retrieved from the database. Each of the
above activities is executed individually
and must be terminated before switching to
another type of activity. The lifetime of
UIF must overlap with DDF, CPS and UTL in
order to support the interactive dialouge
with the user. The lifetime of DBM must
overlap with CPS in order to provide for
the dialouge-oriented access to the
database.

The above considerations resulted in the
decision to implement the UIF, DBM, CPS,
DDF, and UTL objects as processes (tasks).
Another important argument for having them
as processes was that the internal state
of each of those objects is reasonably
large and longlived to justify providing a
separate "state manager" for it. Moreover,
overlapping of the lifetimes of DBM and
UIF offered a potential increase of
efficiency (e.g. DBM starts to execute an
incomplete transaction while UIF is still
interacting with the user).

Cooperation between concurrent processes
is controlled through a protocol. Our goal
was to define a protocol which can be used
to change the system behaviour from a
corutine like mode (where process
switching is completely under the control
of the protocol) to a concurrent mode,
where the protocol designates a number of
"ready" processes which then progress with
the unspecified speeds. This approach

proved to be very useful during
implementation, facilitating testing,
incremental development and deferred
efficiency optimization.

The protocol was specified using the
communicating finite state machine
approach (CFSM) [Brand 1983] and verified
by a semiformal analysis. The important
design goal was to separate the
communication task of a process from its
internal state changes which had very weak
relation to the concurrency control
issues. It was achieved by encapsulating
the internal state transitions in a
separate module which interfaces to the
process through two generic procedures:

- an initialization procedure, and
- an iterator procedure which produces a
  transition to the successor internal
  state.

The information flow from the internal
state changes to the protocol is highly
restricted and is effected by means of the
returned parameters from the above
procedures. The protocol concentrates on
the following issues:

- precess creation,
- establishing interprocess communication
  links,
- maintaining proper synchronization with
  regard to common global resources
  (concurrent file access),
- graceful termination of processes in
  exit and error situations.

The communication takes place along the
communication links which are established
after the corresponding tasks are created.
The synchronous message passing is assumed
(i.e. the exchange of a message is
possible only if the two communicating
parties are willing to communicate;
otherwise the task which is willing to
communicate is delayed).

THE PROTOCOL

The following means are used to implement
the protocol:
- implemenation of each major system
  object as a process;
- separation of the communication and
  internal state transition aspects of an
  object;
- encapsulation of interprocess
  communication services through a common
  interprocess communication (IPC) module
  which offers services to establish
  communication links, sending and
  receiving of messages, and waiting for
  single of multiple source messages
  (nondeterministic receive). This module
  hides completely the services of the
  underlying operating system.

The following messages are exchanged
between processes which run the protocol:

ABRT - "abort me" request from the
        sender. The message is sent if the
        process can not continue normally
        (e.g. can not get enough resources);
CLCI - the UIF process returns the user
        choice on the current menu;
CTSK - the creator process orders the
        receiving process to create
        communication links to some other

process;
DONE — the sending process confirms that the required service is done;
EXIT — the sender communicates a request for termination;
FINI — the creator process orders its child to terminate;
MENU — characterizes the menu to be displayed by the receiving process;
PCNT — the sending process confirms that the required service has been partially realized and some partial results are available;
REDY — means that the sending process is ready to execute;
STRT — the creator process orders its child to start normal execution.

An example specification of the protocol run by the PDB and UIF processes is given below. L1,L2,L3 are the communication links (two way mailboxes) to send/receive messages. For PDB the links are connected as follows:
L1 — to the UIF process,
L2 — to a "user agent process" (DDF,CPS or UTL),
L3 — is used only if L2 is connected to CPS. Then it links to the DBM process,
and for UIF:
L1 — to the PDB process,
L2 — to a "user agent" (DDF,CPS or UTL).

The specification is given in a tabular form. The rightmost column gives state labels. The execution starts in state S1. The second column specifies the communication link(s) from which the process receives being in a given state. The third column specifies logical condition and the fourth column the action taken when a message has been received and the corresponding condition is met. The last column specifies the next state assumed by the process. T denotes the "process termination" state.

THE IPC MECHANISM

The interprocess communication mechanism is based on synchronous message passing. The mechanism has been built above the VMS operating system, using the operating system services (mainly from the $QIO family). From the point of view of the rest of the application, the whole mechanism is represented by the IPC module. The specification of the module is given in Fig.2.

In the DEFINITIONS section the module introduces three types. MES_TYPE defines the data structure for messages exchanged between processes. It is a variant record which enables interpretation of the message according to the MSGTYP type, or alternatively, treating it as an unstructured stream of characters. MSGTYP is a variant record listing possible structures of messages and is defined by the SYS definition module inherited by IPC. The MBX_TYPE defines the data structure for a one way communication mailbox.

The IPC module has no internal state. All data structures which are operated on by the procedures exported by the module must be declared outside the IPC module (i.e. must be declared by the IPC users).

The operations (procedures) exported by the module provide for the following types of usage of the module:

(1) mailbox manipulation, CREATE_MBX, DELETE_MBX;

(2) one-to-one synchronous communication which is achieved by calling SEND by the sending side and calling WAIT by the receiving side;

(3) many-to-one synchronous communication which is achieved by calling SEND by the sending sides and obeying the protocol defined in the ASSUMPTIONS section by the receiving side.

CONCLUSION

The PDBS system has been developed within the timescale and within the budget alotted. Although this success results from the whole outfit of factors (c.f. [Gliss and Gorski 1989]), one of the outstanding reasons is (in our feeling) the presented above approach to concurrency. Proper modularization led to the situation where only few modules "know" about the idiosyncrasies of the underlying operating system. In particular, the whole concurrency control is based on the well founded concepts offered by the IPC module. A simple protocol is used to synchronize processes. This protocol is highly separated from functionality provided by the processes. For instance, the NEXTSCREEN call in the UIF process represents the whole menu presented to the user; the only result of interest is CNCL — which means that the user pressed the "cancel" key on his keyboard. This separation provided for separate testing of the protocol and the functional part of the processes. The protocol itself has been developed in two steps: the coroutine-like version where at any moment only one process is designated as "ready", and the concurrent version which allows more than one "ready" process. Switching between the two versions was achieved by simple and well defined modification and was excercised during system testing and debugging.

REFERENCES

[Booch 1986] G. Booch: Object-oriented development. IEEE Transactions Software Eng., vol. SE-12, no. 2, 1986, pp. 211-221.

[Brand 1983] D. Brand and P. Zafiropulo: On communicating finite state machines. J. Ass. Comput. Mach., vol 30, 1983, pp. 361-371.

[Gliss 1988] B. Gliss: A Guide To Using the PDBS Database System. Internal Report, Max-Planck-Institut, Stuttgart, 1988.

[Gliss and Gorski 1989] B. Gliss and J. Gorski: A Software Engineering Case Study: Using VAX Pascal Extensions in Object Oriented Approach. Submitted for publication.

[VMS Pascal 1987] VAX Pascal Ref. Manual, Digital Equipment Corp., Feb. 1987.

The system manager process PDB.

STATE	LINK	CONDITION	ACTION	NEXT
S1	L1	initialization successful	–	S2
		initialization not successful	–	T
S2	L1	L1 = REDY	L1(MENU)	S3
S3	L1	L1 = CLCI	L1(CTSK)	S4
		L1 = EXIT	L1(FINI)	T
S4	L1	L1 = ABRT	L1(FINI)	T
		L1 = REDY	create user agent process	S5
S5	L2	L2 = ABRT	L2(FINI);L1(FINI)	T
		L2 = REDY		
		agent=CPS	create DBM process	S6
		agent=DDF,UTL	L1(STRT);L2(STRT)	S8
S6	L3	L3 = REDY	L3(CTSK)	S7
S7	L3	L3 = ABRT	L3(FINI);L2(FINI);L1(FINI)	T
		L3 = REDY	L1(STRT);L2(STRT);L3(STRT)	S8
S8	L2	L2 = EXIT	L2(FINI);L3(FINI);L1(MENU)	S3
T	–	–	terminate	

The user interface process UIF.

STATE	LINK	CONDITION	ACTION	NEXT
S1	–	initialization successful	L1(REDY)	S2
		initialization not successful	–	T
S2	L1	L1 = MENU(SSSC)		
		NEXTSCREEN=CNCL	L1(ABRT)	S3
		NEXTSCREEN≠CNCL	L1(CLCI)	S3
S3	L1	L1 = FINI	–	T
		L1 = CTSK	create task	
		creation not OK	L1(ABRT)	S4
		creation OK	L1(REDY)	S4
S4	L1	L1 = FINI	–	T
		L1 = STRT	–	S5
S5	L2	L2 = EXIT	–	S2
		L2 = MENU		
		NEXTSCREEN=CNCL	L1(ABRT)	S5
		NEXTSCREEN≠CNCL	L1(CLCI)	S5
T	–	–	terminate	

Fig. 1. The protocol definition for the PDB and UIF processes.

```
[INHERIT('SYS$LIBRARY:STARLET','SYS','LIB'),
ENVIRONMENT('IPC')]
{ SPECIFICATION } MODULE IPC;

{ 8-10-87 JG }
{------------------------------------------------------------------------}
{ DESCRIPTION: The module offers a set of procedures for                 }
{              interprocess communication.                               }
{------------------------------------------------------------------------}
{ DEFINITIONS:                                                           }
{-----------------}
     TYPE

                     { container type for messages exchanged through  }
                     { mailboxes                                       }
             MES_TYPE=RECORD
                     CASE BOOLEAN OF
                        TRUE :(RESERVED:INTEGER; MSGSTRUC:MSGTYP);
                        FALSE:(MSGTRANS:VARYING [MESSAGE_LEN] OF CHAR);
                     END;


                     { standard data structure for communication      }
                     { through VMS $QIO function                       }
             IO_BLOCK=RECORD
                         IO_STAT,COUNT:UWORD;
                         DEV_INFO:INTEGER;
                     END;


                     { mailbox representation                          }
             MBX_TYPE=RECORD
                         NAME:MBX_NTYPE;
                         CHAN:UWORD;         { channel number          }
                         EFN:UNSIGNED;       { event flag number       }
                         IOSB:IO_BLOCK;      { IO status block         }
                         DATA:MES_TYPE;      { message container       }
                     END;
{------------------------------------------------------------------------}
{ STATE:    none                                                         }
{------------------------------------------------------------------------}
{ OPERATIONS:                                                            }
{------------------}

     [EXTERNAL]FUNCTION CREATE_MBX(VAR MBX:MBX_TYPE;
        NAME:MBX_NTYPE;EVENT:UNSIGNED):BOOLEAN;
     EXTERN;
     { DESCRIPTION: Creates a communication mailbox.                     }
     { PARAMETERS:                                                       }
     {         MBX - a data structure which represents the              }
     {               mailbox.                                           }
     {         NAME - a name given to the mailbox.                      }
     {         EVENT - an event flag number for the event              }
     {                 used to synchronize the process while           }
     {                 receiving from the mailbox.                      }
     { RETURNS: TRUE - if mailbox created                               }
     {          FALSE- if mailbox not created (any condition           }
     {                   different than "SUCCESS" signalled            }
     {                   by the VMS $CREMBX service routine            }
     {                   causes this function to fail.                  }
     { ASSUMPTION: The event parameter should be set 0 for all         }
     {         mailboxes except those which will be used in            }
     {         parallel wait (PAR_WAIT). Such mailboxes should         }
     {         be associated with an unique (local) event number       }

     {         i.e. an integer from the interval 1..31 .                }
     { EFFECT: Initializes the MBX data structure and associates       }
     {         it with the 'NAME'  mailbox                             }
     {         If the VMS exception is detected while creating         }
     {         the mailbox, a message which indicates the IPC          }
     {         module is sent to the terminal and the call            }
     {         returnes "FALSE".                                        }
{------------------------------------------------------------------------}

     [EXTERNAL]PROCEDURE DELETE_MBX(VAR MBX:MBX_TYPE);
     EXTERN;
     { DESCRIPTION: Deletes the MBX mailbox                             }
     { PARAMETERS:                                                      }
     {         MBX - the mailbox to be deleted                         }
```

Fig. 2. (continued on the next page) The specification of IPC module.

```
{ EFFECT: Deletes the MBX mailbox and releases all          }
{          operating system resources allocated to it.      }
{          If the VMS exception is detected while deleting  }
{          the mailbox, a message which indicates the IPC   }
{          module is sent to the terminal and the calling   }
{          process is stopped (LIB$STOP).                    }
{------------------------------------------------------------}

    [EXTERNAL]PROCEDURE SEND(VAR MBX:MBX_TYPE);
    EXTERN;
{ DESCRIPTION: Sends a message synchronously                }
{ PARAMETERS:                                               }
{          MBX - a data structure which represents a mailbox }
{              MBX.DATA contains a message to be sent.      }
{ EFFECT: The calling process is delayed until a message    }
{          is delivered through MBX to the process which    }
{          receives from the same channel. The message      }
{          is returned  in MBX.DATA.                        }
{          Any o.s. exceptions received by SEND cause that  }
{          a message:                                       }
{          'IPC: SEND ERROR ON',MBX.NAME                    }
{          is sent to the terminal                          }
{------------------------------------------------------------}

    [EXTERNAL]PROCEDURE WAIT(VAR MBX:MBX_TYPE);
    EXTERN;
{ DESCRIPTION: synchronous wait on a mailbox                }
{ PARAMETERS:                                               }
{          MBX - a data structure that represents a mailbox. }
{ EFFECT: The calling process is delayed until a message    }
{          is received through MBX. The message is returned }
{          in MBX.DATA.                                     }
{          Any o.s. exceptions received by SEND cause that  }
{          a message:                                       }
{          'IPC: WAIT ERROR ON',MBX.NAME                    }
{          is sent to the terminal                          }
{------------------------------------------------------------}

    [EXTERNAL]FUNCTION RECEIVE(VAR MBX:MBX_TYPE):BOOLEAN;
    EXTERN;
{ DESCRIPTION: activates asynchronous receiving from a mailbox }
{ PARAMETERS:                                               }
{          MBX - a data structure that represents a mailbox. }
{ EFFECT: Asynchronous receiving from MBX is initiated.     }
{          The call causes no delay of the calling process. }
{          The function returns TRUE if the call was successful }
{          i.e. MBX has been initiated. Otherwise the       }
{          function returns FALSE.                          }
{ ASSUMPTION: The function should be used together with     }
{          PAR_WAIT to implement parallel wait on mailboxes. }

{------------------------------------------------------------}

    [EXTERNAL]PROCEDURE PAR_WAIT(VAR MBX1,MBX2:MBX_TYPE;
          VAR DATA1,DATA2:MES_TYPE;VAR MBX1RES,MBX2RES:BOOLEAN);
    EXTERN;
{ DESCRIPTION: parallel receiving from two mailboxes.       }
{ PARAMETERS:                                               }
{      MBX1,MBX2 - mailboxes from which messages are to be  }
{                  received                                 }
{      DATA1,DATA2 - containers for messages received.      }
{      MBX1RES,MBX2RES - boolean indicators which indicate  }
{          which mailbox received a message (i.e. if MBXiRES is }
{          TRUE then the I-th mailbox received a message and }
{          the message is returned in DATAi).               }
{ EFFECT: The procedure allows for parallel wait on two     }
{          mailboxes. It gives priority to MBX1 (i.e. if messages }
{          arrive thruogh MBX1 with high frequency, MBX2 can be }
{          locked out. Only one message is received by PAR_WAIT. }
{          If no messages are available from MBX1 and MBX2, the }
{          calling process is delayed.                      }
{------------------------------------------------------------}
{ ASSUMPTIONS:                                              }
{----------------------------------}
{    To receive from two mailboxex in parallel, the following }
{    protocol must be obeyed:                               }
{    (1) MBX1 and MBX2 must have EFN fields set (the        }
{        flag numbers)                                      }
{    (2) RECEIVE(MBX1) and RECEIVE(MBX2) must be successfully }
{        called once, at process initialization            }
{    (3) then, PAR_WAIT on MBX1 and MBX2 can be repeatedly called }
{------------------------------------------------------------}
END.{IPC}
{------------------------------------------------------------}
```

ON PRIORITY-BASED SYNCHRONIZATION PROTOCOLS FOR DISTRIBUTED REAL-TIME DATABASE SYSTEMS

Sang H. Son

*Department of Computer Science, University of Virginia, Charlottesville,
VA 22903, USA*

Abstract: Real-time database systems must maintain consistency while minimizing the number of transactions that miss the deadline. To satisfy both the consistency and real-time constraints, there is the need to integrate synchronization protocols with real-time priority scheduling protocols. This paper describes a prototyping environment for investigating distributed real-time database systems, and its use for performance evaluation of priority-based scheduling protocols for real-time database systems.

Keywords: distributed database, real-time, prototyping, synchronization, transaction, priority.

1. Introduction

As computers are becoming essential part of real-time systems, *real-time computing* is emerging as an important discipline in computer science and engineering [Shin87]. The growing importance of real-time computing in a large number of applications, such as aerospace and defense systems, industrial automation, and nuclear reactor control, has resulted in an increased research effort in this area. Since any kind of computing needs to access data, methods for designing and implementing database systems that satisfy the requirement of timing constraints in collecting, updating, and retrieving data play an important role in the success of real-time systems.

Researchers working on developing new real-time systems based on distributed system architecture have found out that database managers are assuming much greater importance in real-time systems [Son88]. One of the characteristics of current database managers is that they do not schedule their transactions to meet response requirements and they commonly lock data tables indiscriminately to assure database consistency. Locks and time-driven scheduling are basically incompatible. Low priority transactions can and will block higher priority transactions leading to response requirement failures. New techniques are required to manage database consistency which are compatible with time-driven scheduling and the essential system response predictability/analyzability it brings. One of the primary reasons for the difficulty in successfully developing and evaluating a distributed database system is that it takes a long time to develop a system, and evaluation is complicated because it involves a large number of system parameters that may change dynamically.

A prototyping technique can be applied effectively to the evaluation of control mechanisms for distributed database systems. A *database prototyping tool* is a software package that supports the investigation of the properties of a database control techniques in an environment other than that of the target database system. The advantages of an environment that provides prototyping tools are obvious. First, it is cost effective. If experiments for a twenty-node distributed database system can be executed in a software environment, it is not necessary to purchase a twenty-node distributed system, reducing the cost of evaluating design alternatives. Second, design alternatives can be evaluated in a uniform environment with the same system parameters, making a fair comparison. Finally, as technology changes, the environment need only be updated to provide researchers with the ability to perform new experiments.

A prototyping environment can reduce the time of evaluating new technologies and design alternatives. From our past experience, we assume that a relatively small portion of a typical database system's code is affected by changes in specific control mechanisms, while the majority of code deals with intrinsic problems, such as file management. Thus, by properly isolating technology-dependent portions of a database system using modular programming techniques, we can implement and evaluate design alternatives very rapidly. Although there exist tools for system development and analysis, few prototyping tools exist for distributed database experimentation. Especially if the system designer must deal with message-passing protocols and timing constraints, it is essential to have an appropriate prototyping environment for success in the design and analysis tasks.

This paper describes a message-based approach to prototyping study of distributed real-time database systems, and presents a prototyping software implemented for a series of experimentation to evaluate priority-based synchronization algorithms.

This work was supported in part by the Office of Naval Research under contract number N00014-88-K-0245 and by the Federal Systems Division of IBM Corporation under University Agreement WG-249153.

2. Structure of the Prototyping Environment

For a prototyping tool for distributed database systems to be effective, appropriate operating system support is mandatory. Database control mechanisms need to be integrated with the operating system, because the correct functioning of control algorithms depends on the services of the underlying operating system; therefore, an integrated design reduces the significant overhead of a layered approach during execution.

Although an integrated approach is desirable, the system needs to support flexibility which may not be possible in an integrated approach. In this regard, the concept of developing a library of modules with different performance and reliability characteristics for an operating system as well as database control functions seems promising. Our prototyping environment follows this approach [Cook87, Son88b]. It is designed as a modular, message-passing system to support easy extensions and modifications. An instance of the prototyping environment can manage any number of virtual sites specified by the user. Modules that implement transaction processing are decomposed into several server processes, and they communicate among themselves through ports. The clean interface between server processes simplifies incorporating new algorithms and facilities into the prototyping environment, or testing alternate implementations of algorithms.

User Interface (UI) is a front-end invoked when the prototyping environment begins. UI is menu-driven, and designed to be flexible in allowing users to experiment various configurations with different system parameters. A user can specify the following:

- system configuration: number of sites and the number of server processes at each site.
- database configuration: database at each site with user defined structure, size, granularity, and levels of replication.
- load characteristics: number of transactions to be executed, size of their read-sets and write-sets, transaction types (read-only or update) and their priorities, and the mean interarrival time of transactions.
- concurrency control: locking, timestamp ordering, and priority-based.

UI initiates the Configuration Manager (CM) which initializes necessary data structures for transaction processing based on user specification. CM invokes the Transaction Generator at an appropriate time interval to generate the next transaction to form a Poisson process of transaction arrival.

Transaction execution consists of read and write operations. Each read or write operation is preceded by an access request sent to the Resource Manager, which maintains the local database at each site. Each transaction is assigned to the Transaction Manager (TM). TM issues service requests on behalf of the transaction and reacts appropriately to the request replies.

The Performance Monitor interacts with the transaction managers to record, priority/timestamp and read/write data set for each transaction, time when each event occurred, statistics for each transaction and cpu hold interval in each node. The statistics for a transaction includes arrival time, start time, total processing time, blocked interval, whether deadline was missed or not, and number of aborts.

3. Priority-Based Synchronization

In a real-time database system, synchronization protocols must not only maintain the consistency constraints of the database but also satisfy the timing requirements of the transactions accessing the database. To satisfy both the consistency and real-time constraints, there is the need to integrate synchronization protocols with real-time priority scheduling protocols. A major source of problems in integrating the two protocols is the lack of coordination in the development of synchronization protocols and real-time priority scheduling protocols. Due to the effect of blocking in lock-based synchronization protocols, a direct application of a real-time scheduling algorithm to transactions may result in a condition known as *priority inversion*. Priority inversion is said to occur when a higher priority process is forced to wait for the execution of a lower priority process for an indefinite period of time. When the transactions of two processes attempt to access the same data object, the access must be serialized to maintain consistency. If the transaction of the higher priority process gains access first, then the proper priority order is maintained; however, if the transaction of the lower priority gains access first and then the higher priority transaction requests access to the data object, this higher priority process will be blocked until the lower priority transaction completes its access to the data object. Priority inversion is inevitable in transaction systems. However, to achieve a high degree of schedulability in real-time applications, priority inversion must be minimized. This is illustrated by the following example.

Example: Suppose T_1, T_2, and T_3 are three transactions arranged in descending order of priority with T_1 having the highest priority. Assume that T_1 and T_3 access the same data object O_i. Suppose that at time t_1 transaction T_3 obtains a lock on O_i. During the execution of T_3, the high priority transaction T_1 arrives, preempts T_3 and later attempts to access the object O_i. Transaction T_1 will be blocked, since O_i is already locked. We would expect that T_1, being the highest priority transaction, will be blocked no longer than the time for transaction T_3 to complete and unlock O_i. However, the duration of blocking may, in fact, be unpredictable. This is because transaction T_3 can be blocked by the intermediate priority transaction T_2 that does not need to access O_i. The blocking of T_3, and hence that of T_1, will continue until T_2 and any other pending intermediate priority level transactions are completed.

The blocking duration in the example above can be arbitrarily long. This situation can be partially remedied if transactions are not allowed to be preempted; however, this solution is only appropriate for very short transactions, because it creates unnecessary blocking. For instance, once a long low priority transaction starts execution, a high priority transaction not requiring access to the same set of data objects may be needlessly blocked.

An approach to this problem, based on the notion of *priority inheritance*, has been proposed [Sha87]. The basic idea of priority inheritance is that when a transaction T of a process blocks higher priority processes, it executes at the highest priority of all the transactions blocked by T. This simple idea of priority inheritance reduces the blocking time of a higher priority transaction. However, this is inadequate because the blocking duration for a transaction, though bounded, can still be substantial due to the potential *chain of blocking*. For instance, suppose that transaction T_1 needs to sequentially access objects O_1 and O_2. Also suppose that T_2 preempts T_3 which has already locked O_2. Then, T_2 locks O_1. Transaction T_1 arrives at this instant and finds that the objects O_1 and O_2 have been respectively locked by the lower priority transactions T_2 and T_3. As a result, T_1 would be blocked for the duration of two transactions, once to wait for T_2 to release O_1 and again to wait for T_3 to release O_2. Thus a chain of blocking can be formed.

Several methods to combat this inadequacy are under investigation. The *priority ceiling protocol* is one of such methods being investigated at the Carnegie-Mellon University [Sha88]. It tries to achieve not only minimizing the blocking time of a transaction to at most one lower priority transaction execution time, but also preventing the formation of deadlocks. The priority ceiling protocol has been implemented in our real-time database system and compared with other synchronization protocols using the prototyping environment.

Using the prototyping tool, we have been evaluating the priority ceiling protocol and investigating technical issues associated with priority-based scheduling protocols. One of the issues we are studying is the comparison of the priority ceiling protocol with other design alternatives. In our experiments, all transactions are assumed to be *hard* in the sense that there will be no value in completing a transaction after its deadline. Transactions that miss the deadline are aborted, and disappear from the system immediately with some abort cost.

4. Priority Ceiling Protocol

The priority ceiling protocol is premised on systems with a fixed priority scheme. The protocol consists of two mechanisms: *priority inheritance* and *priority ceiling*. We already have explained the priority inheritance mechanism. In the priority ceiling mechanism, a priority ceiling is defined for every data object as the priority of the highest priority transaction which may access the data object. A transaction is allowed to access the data object only if its priority is higher than the priority ceilings of all data objects currently being accessed by some transaction in the system. With the combination of these two mechanisms, it has been shown that in the worst case, each transaction has to wait for at most one lower priority transaction in its execution, and no deadlock will ever occur [Sha88]. In the next example, we show how transactions are scheduled under the priority ceiling protocol.

Example: Consider the same situation as in the previous example. According to the protocol, the priority ceiling of O_i is the priority of T_1. When T_2 tries to access a data object, it is blocked because its priority is not higher than the priority ceiling of O_i. Therefore T_1 will be blocked only once by T_3 to enter O_i, regardless of the number of data objects it may access.

The ceiling manager implements the priority ceiling algorithm in the prototyping environment. The lock on a data object can either be a read-lock or a write-lock. The write-priority ceiling of a data object is defined as the priority of the highest priority transaction that may write into this object, and absolute-priority ceiling is defined as the priority of the highest priority transaction that may read or write the data object. When a data object is write-locked (read-locked), the rw-priority ceiling of this data object is defined to be equal to the absolute (write) priority ceiling.

When a transaction attempts to lock a data object, the transaction's priority is compared with the highest rw-priority ceiling of all data items currently locked by other transactions. If the priority of the transaction is not higher than the rw-priority ceiling, it will be denied. Otherwise, it is granted the lock. In the denied case, the priority inheritance is performed in order to overcome the problem of uncontrolled priority inversion.

Under this protocol, it is not necessary to check for the possibility of read-write conflicts. For instance, when a data object is write-locked by a transaction, the rw-priority ceiling is equal to the highest priority transaction that can access it. Hence, the protocol will block a higher priority transaction that may write or read it. On the other hand, when the data object is read-locked, the rw-priority ceiling is equal to the highest priority transaction that may write it. Hence, a transaction that attempts to write it will have a priority no higher than the rw-priority ceiling and will be blocked. Only the transaction that read it and have priority higher than the rw-priority ceiling will be allowed to read-lock it, since read-locks are compatible.

5. Performance Evaluation

Various statistics have been collected for comparing the performance of the priority-ceiling protocol with other synchronization control algorithms. Transaction are generated with exponentially distributed interarrival times, and the data objects updated by a transaction are chosen uniformly from the database. A transaction has an execution profile which alternates data access requests with equal computation requests, and some processing requirement for termination (either commit or abort). Thus the total processing time of a transaction is directly related to the number of data objects accessed. Due to space considerations, we cannot present all our results but have selected the graphs which best illustrate the difference and performance of the algorithms. For example, we have omitted the results of an experiment that varied the size of the database, and thus the number of conflicts, because they only confirm and not increase the knowledge yielded by other experiments.

For each experiment and for each algorithm tested, we collected performance statistics and averaged over the 10 runs. The percentage of deadline-missing transactions is calculated with the following equation: $\%missed = 100 *$ (number of deadline-missing transactions / number of transactions processed). A transaction is processed if either it executes completely or it is aborted. We assume that all the transactions are *hard* in the sense that there will be no value for completing the transaction after its deadline. Transactions that miss the deadline are aborted, and disappeared from the system immediately with some abort cost. We have used the transaction size (the number of data objects a transaction needs to access) as one of the key variables in the experiments. It varies from a small fraction up to a relatively large portion (10%) of the database so that conflict would occur frequently. The high conflict rate allows synchronization protocols to play a significant role in the system performance. We choose the arrival rate so that protocols are tested in a heavily loaded rather than lightly loaded system. It is because for designing real-time systems, one must consider high load situations. Even though they may not arise frequently, one would like to have a system that misses as few deadlines as possible when such peaks occur. In other words, when a crisis occurs and the database system is under pressure is precisely when making a few extra deadlines could be most important [Abb88].

We normalize the transaction throughput in records accessed per second for successful transactions, not in transactions per second, in order to account for the fact that bigger transactions need more database processing. The normalization rate is obtained by multiplying the transaction completion rate (transactions/second) by the transaction size (database records accessed/transaction). In Figure 1, the throughput of the priority-ceiling protocol (C), the two-phase locking protocol with priority mode (P), and the two-phase locking protocol without priority mode (L), is shown for transactions of different sizes with balanced workload and I/O bound workload.

As the transaction size increases, there is little impact on the throughput of priority-ceiling protocol over the range of transaction sizes and over the workload type shown in Figure 1. This is because in priority-ceiling protocol, the conflict rate is determined by ceiling blocking rather than direct blocking, and the frequency of ceiling blocking is not sensitive to the transaction size.

However, the performance of the two-phase locking protocol with or without priority degrades very rapidly. This phenomenon is more clear as the transaction workload is closer to I/O bound, since there are few conflicts for the small transactions in the two-phase locking protocol, and the concurrency is fully achieved in the assumption of parallel I/O processing. Poor performance of the two-phase locking protocol for bigger transactions is due to the high conflict rate.

Since I/O cost is one of the key parameters in determining performance, we have investigated an approach to improve system performance by performing I/O operation before locking. This is called the *intention I/O*. In the intention mode of I/O operation, the system pre-fetches data objects that are in the access lists of transactions submitted, without locking them. This approach will reduce the locking time of data objects, resulting in higher throughput. As shown in Figure 2, intention I/O improves throughput of both the two-phase locking and the ceiling protocol. However, improvement in the ceiling protocol is much more significant. This is because the frequency of ceiling blocking is very sensitive to the duration of data object locking in the system.

Another important performance statistics is the percentage of deadline missing transactions, since the synchronization protocol in real-time database systems must satisfy the timing constraint of individual transaction. In our experiments, each transaction's deadline is set to proportional to its size and system workload (number of transactions), and the transaction with the earliest deadline is assigned the highest priority. As shown in Figure 3, the percentage of deadline missing transactions increases sharply for the two-phase locking protocol as the transaction size increases. A sharp rise was expected, since the probability of deadlocks would go up with the fourth power of the transaction size [Gray81]. However, the percentage of deadline missing transactions increases much slowly as the transaction size increases in the priority-ceiling protocol, since there is no deadlock in priority-ceiling protocol and the response time is proportional to the transaction size and the priority ranking.

6. Conclusions

Prototyping large software systems is not a new approach. However, methodologies for developing a prototyping environment for distributed database systems have not been investigated in depth in spite of its potential benefits. In this paper, we have presented a prototyping environment that has been developed based on a message-based approach with modular building blocks. Although the complexity of a distributed database system makes prototyping difficult, the implementation has proven satisfactory for experimentation of design choices, different database techniques and protocols, and even an integrated evaluation of database systems. It supports a very flexible user interface to allow a wide range of system configurations and workload characteristics. Expressive power and performance evaluation capability of our prototyping environment has been demonstrated by implementing a distributed real-time database system and investigating its performance characteristics.

There are many technical issues associated with priority-based synchronization protocols that need further investigation. For example, the analytic study of the priority ceiling protocol provides an interesting observation that the use of read and write semantics of a lock may lead to worse performance in terms of schedulability than the use of exclusive semantics of a lock. This means that the *read* semantics of a lock cannot be used to allow several readers to hold the lock on the data object, and the ownership of locks must be mutually exclusive. Is it necessarily true? We are investigating this and other related issues using the prototyping environment.

References

[Abb88] Abbott, R. and H. Garcia-Molina, "Scheduling Real-Time Transactions: A Performance Study," *VLDB Conference,* Sept. 1988, pp 1-12.

[Cook87] Cook, R. and S. H. Son, "The StarLite Project," *Fourth IEEE Workshop on Real-Time Operating Systems,* Cambridge, Massachusetts, July 1987, 139-141.

[Gray81] Gray, J. et al., "A Straw Man Analysis of Probability of Waiting and Deadlock," *IBM Research Report,* RJ 3066, 1981.

[Sha87] Sha, L., R. Rajkumar, and J. Lehoczky, "Priority Inheritance Protocol: An Approach to Real-Time Synchronization," *Technical Report,* Computer Science Dept., Carnegie-Mellon University, 1987.

[Sha88] Sha, L., R. Rajkumar, and J. Lehoczky, "Concurrency Control for Distributed Real-Time Databases," *ACM SIGMOD Record 17,* 1, Special Issue on Real-Time Database Systems, March 1988.

[Shin87] Shin, K. G., "Introduction to the Special Issue on Real-Time Systems," *IEEE Trans. on Computers,* Aug. 1987, 901-902.

[Son88] Son, S. H., "Real-Time Database Systems: Issues and Approaches," *ACM SIGMOD Record 17,* 1, Special Issue on Real-Time Database Systems, March 1988.

[Son88b] Son, S. H., "A Message-Based Approach to Distributed Database Prototyping," *Fifth IEEE Workshop on Real-Time Software and Operating Systems,* Washington, DC, May 1988, 71-74.

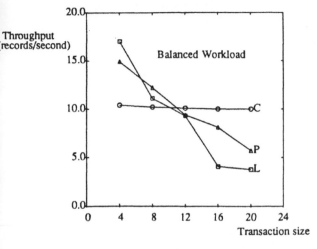

a) balanced workload transaction

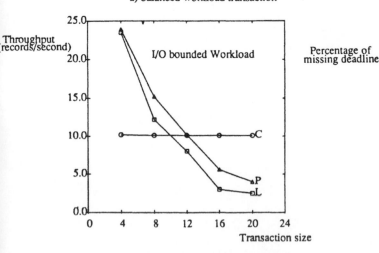

b) I/O bounded workload transaction

Fig. 1 Transaction Throughput.

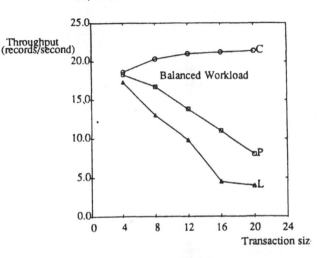

Fig. 2 Transaction Throughput with Intention I/

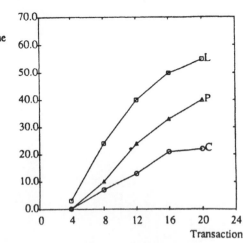

Fig. 3 Percentage of Missing Deadline .

C: priority_ceiling protocol
P: 2-phase locking protocol with priority mode
L: 2-phase locking protocol without priority mode

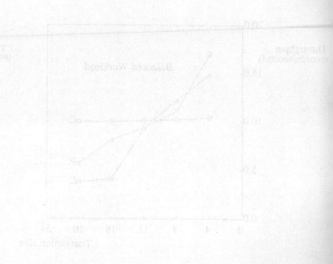

a) Balanced workload transaction.

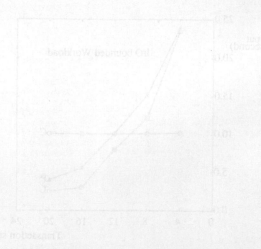

b) I/O bounced workload transaction.

Fig. 1 Transaction Throughput.

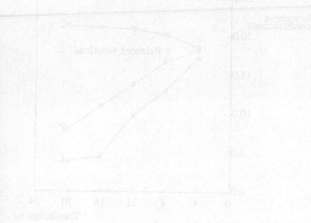

Fig. 2 Transaction Throughput with Iteration in

Fig. 4 Percentage of Missing Deadline

MODELLING AND VERIFICATION OF REAL-TIME SOFTWARE USING INTERPRETED PETRI NETS

F. Arendt and B. Klühe

Academy of Sciences of the GDR,
Central Institute of Cybernetics and Information Processes, P.O. Box 1298,
Kurstraße 33, Berlin, 1086, GDR

Abstract. Increasing the efficiency of the software development and guaranteeing the software correctness has become a main problem in real-time programming. This problem connected with the parallel operating mode in real-time systems and general approaches for its solution are discussed. It can only be solved with the help of mathematical models suitable for a compact and transparent software description as well as for a verification of the dynamic behaviour of parallel programs. In this paper a method is presented for modelling and verification using interpreted Petri nets. An adapted <u>S</u>oftware <u>I</u>nterpreted <u>N</u>et (SIN) is introduced. It allows firstly the description of the software on a problem-oriented level and secondly the <u>complete verification</u> on the basis of Petri net theory. The principles of modelling and verification are demonstrated. Properties of the program can be proved by analysing the corresponding properties of the SIN. The connection between results of the analysis on SIN level and the properties of the modelled software are demonstrated by an example. A software tool for realizing the method described is under development. Its structure is shown.

Keywords. High level languages; Petri nets; Modeling; Program testing; Real time computer systems; Software tools.

THE VERIFICATION PROBLEM

The development in the field of micro-computers and their application is characterized by an increasing efficiency of computers leading to an increasing complexity of application tasks and vice versa. More and more distributed computer-systems are used for the solution of advanced technical problems.

The software, mostly under control of a real-time operating system, consists of parallel processes synchronizing themselves in well-defined states by communication. Because of the large number of reachable program states as a consequence of the parallel operating mode the testing used in sequential programming is insufficient for a proof of the correctness of parallel programs. Increasing the efficiency of the software development and guaranteeing the software correctness has become a main problem.

The only way out of this situation is the application of mathematical models being suitable for a compact and transparent software description as well as for a complete verification of parallel programs. But until now there do not exist any models being of general use for the solution of this problem.

The methods and tools for the verification of parallel software presently known can be divided in the following complexes:

- design methods,
- environment for development and runtime,
- test methods,
- simulation,
- methods for proof and analysis.

Design methods. Numerous approaches have the intention to develop methods supporting the software design process from its beginning, i.e. the problem analysis, to the end continuously and guaranteeing the fulfilment of all reliability requirements for the system a priori. The idea of such methods is to restrict the set of types of objects used and the set of operations on these objects so that promoted system properties can be guaranteed. On the basis of different models some aspects of the problem of design of correct programs could be solved in the past. But until now there is not known any software tool supporting the software design process in general and excluding software errors completely.

Environment for development and runtime. Modern real-time operating systems and programming languages with elements for the description of parallelism may be very useful to increase software reliability. Real-time operating systems support the communication and synchronization in the system. Using these features of the operating system the programmer is free from the problems of a correct realization of the mechanisms of communication and synchronization but not from the task making correct use of them. Algorithmic programming languages with elements for the description of parallel processes support the specification relativly independent from a definite operating system- and hardware-environment. By this the prerequisites are given for the syntactical and semantical checking by the compiler on principle.

Test methods. Existing errors can be detected possibly using tests. But the proof of the absence of any error by testing is impossible for complex systems of parallel processes because of the great number of system states and the dependency on time of the behaviour of the system as a consequence of the parallel operating mode. Another disadvantage of testing consists in the fact that testing is only possible after realization of the system or its components. The detection of errors in this very late state of the software design process causes high expenses for correction.

Simulation. In comparison to testing methods simulation offers two important advantages:

- the behaviour can be examined before the completion of the system,
- even such systems can be examined, for which testing is impossible for technical reasons or for which it is too time consuming.

But the complete proof of the absence of any error in complex systems is also impossible using simulation (for the same reasons like using test methods).

Methods for proof and analysis.
The complete verification of real-time software is only possible using mathematical methods for modelling, analysis and proof of the properties of the program. They were applied for special small problems. Applications of mathematical proof-methods for the verification of complex real-time systems are unknown (QUIRK, 1985).

MODELLING AND VERIFICATION WITH NETS

In this paper a method is presented for modelling and complete verification using interpreted Petri nets. The Petri net theory offers important advantages to apply it in software design and verification:

- Petri net (PN) is a model incorporating elements for a structural as well as a behavioural description of parallel systems,
- graphical net representation results in a high degree of transparency even for large systems,
- the real process is modelled by the flow of tokens through the net,
- PN can be adapted to a special problem class by problem-oriented interpretation of the net elements,
- there are algorithms for a computer-aided analysis and simulation of PNs for several net types.

But there is the following basic contradiction restricting the application of Petri net theory.

Interpreted Petri nets or high level net models using problem-adapted interpretations of places and transitions or modifications of properties of net elements are suitable for an efficient description of both the control and data flow of programs. These models cannot be analysed for verification directly and its simulation is insufficient for a complete verification.

Uninterpreted Petri nets (PN), i.e. place-transition nets (P/T-nets) $N=(P,T,F,C,W,M)$ consist of:

- set of places P
- set of transitions T
- flow relation F (the elements of F are called arcs)
- capacity function of places C
- weight function of arcs W
- initial marking M.

In addition there is a switching rule for the description of the flow of tokens through the net (causing a change of the marking). The behaviour of P/T-nets can be analysed by the following methods:

- analysis of structural properties and concluding from these to the dynamic behaviour
- computation of net invariants supplying information about net properties
- computation of the reachability graph, i.e. the set of all feasable markings of the net inclusively all changes between these markings. The markings are corresponding with the feasable states of the modelled system.

Analysis of structural properties and invariants is only available in special cases. The reachability graph can always be computed for finite nets. The practical use of this method is only limited by the computation time and memory demand for storing the markings. P/T-nets and similar models can be applied for modelling and analysing of programs but lead to obscure descriptions for comlex problems.

This contradiction could be solved by a computer-aided connection between a problem-oriented (interpreted) net model for problem-description and net-oriented algorithms for analysis.

CONCEPT FOR A SOLUTION OF THE PROBLEM

In the paper the so called Software Interpreted Net (SIN) is introduced. The model allows on the one hand the description of the software on a problem-oriented level and on the other hand the complete verification like an uninterpreted PN after its transformation. Constructs being typical for the investigated problem, e.g. Boolean and Integer operations, are described as SIN structures on the level of a programming language. Problems of the refinement of SIN structures and the transformation of SIN into PN are shown. Conclusions from the results of the analysis on PN level, e.g. the analysis of the dynamic behaviour, concerning the correctness of the modelled software can be drawn.

The method presented here includes the following steps:

- modelling and refinement on SIN level
- transformation of the SIN into a PN model
- analysis of properties of the PN model
- conclusions from the results of the analysis onto the software properties looked for.

THE NET MODEL

The modelling of computer programs is carried out usually on the process-oriented (data flow) level or on the implementation-oriented (control flow) level. The models used mostly are control flow-oriented ones. In general it is also necessary to consider the data flow for the verification explicitly, because the software behaviour is influenced by actual values of the variables. For that reason we introduce the Software Interpreted Net (SIN),i.e. an interpreted PN with elements to represent both the control flow and the data flow. Based on the Petri net notion the SIN (Kluehe,1988) is defined as follows.

The tupel SIN = $(N, A, D, q_T, W_0(D))$ is a Software Interpreted Net, with

$N = (P,T,F,C,W,m_0)$ is a Petri net,
A = DCF u DAF is the set of program-statements,
   DCF   is the set of data condition functions (dcf) containing all data related conditions, which influence the control flow,
   DAF   is the set of data action functions (daf), i.e. all statements processing values of variables,
D the set of program variables
q_T: T ---> DCF x P(DAF) is a mapping assigning one or more statements to each transition,
$W_0(D)$ are the initial values of variables.

A state of the program is given completely in a SIN by a marking m and the values of variables W(D).

The following two kinds of requirements have to be met for switching of a transition t:

- marking related condition:all pre-places of t have to contain enough tokens and it is not allowed to exceed the capacity of the post-places by switching of t,
- data related condition: if t has a dcf as an attribute, then this condition must be true for switching of t.

The result of switching of a transition t is a new marking, corresponding with a new program state mentioned above, and the processing of all daf, which are related to this transition by the mapping $q_T(t)$.

The term reachability graph (RG) is significant for the method shown in this paper to investigate parallel programs. Let be N a net, T its set of transitions and C the set of all markings reachable in N from the initial marking m. Than the reachability graph RG is determined by the set of nodes defined by a bijective mapping on C. Nodes c',c'' ∈ C are connected by a directed edge, if there is a transition

transmitting c' to c'' by switching. That means ways in RG are related to possible switching sequences in N. By this RG gives a complete survey on the processes feasible in N. In other words: complete program behaviour is represented by the related reachability graph.

MODELLING

Figure 1 demonstrates the modelling of a program with the help of SIN. Net constructs for the modelling of common elements of programming languages used for composing the control flow are presented in Kluehe(1988).

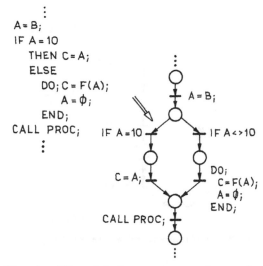

```
          ⋮
     A = B;
     IF A = 10
        THEN C = A;
        ELSE
           DO; C = F(A);
              A = φ;
           END;
     CALL PROC;
          ⋮
```

Fig. 1. SIN model for a program fragment.

REFINEMENT OF A SIN MODEL

The available algorithms for net analysis are suitable for uninterpreted nets only. Because of this the problem has to be modelled on such a SIN level being allowed its transformation to an (uninterpreted) PN. To reach this SIN level and to model all details of the problem being of relevance for verification is the aim of the refinement step by step (Fig. 2).

In this context the mechanism of switching of a transition in a SIN has to be taken into account. The switching of a transition, that means: testing the dcf, flow of tokens from the pre-places to the transition and from the transition to the post-places as well as the realizing of the corresponding dafs is an undivisible operation at all. It follows from this, that single transitions are only capable for modelling of:

- undivisable operations of the program (e.g. semaphore-operations) and
- (sequences) of operations, if their possible interruption is without consequences for the behaviour in each cases.

This connection can be demonstrated by an example (Fig. 3).

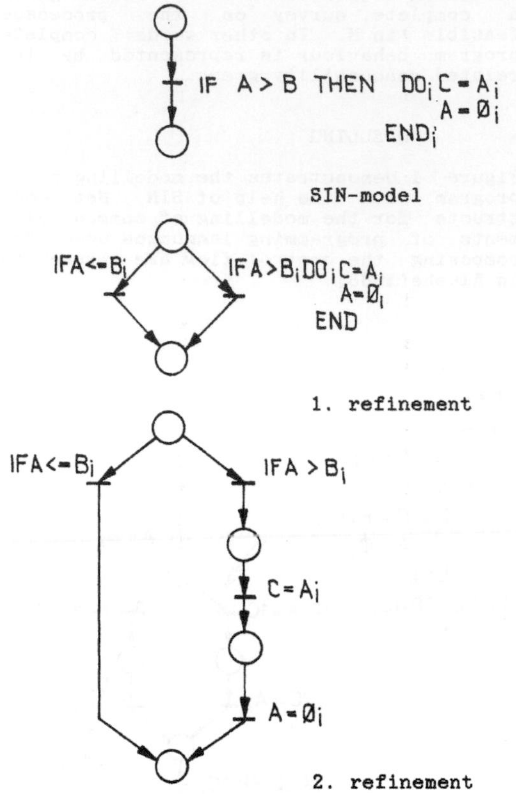

IF A > B THEN DO; C = A;
                            A = Ø;
                      END;

SIN-model

IFA<=B;   IFA>B;DO;C=A;
                              A=Ø;
                          END

1. refinement

IFA<=B;   IFA > B;

                          C = A;

                          A = Ø;

2. refinement

Fig. 2.   Refinement of a transition

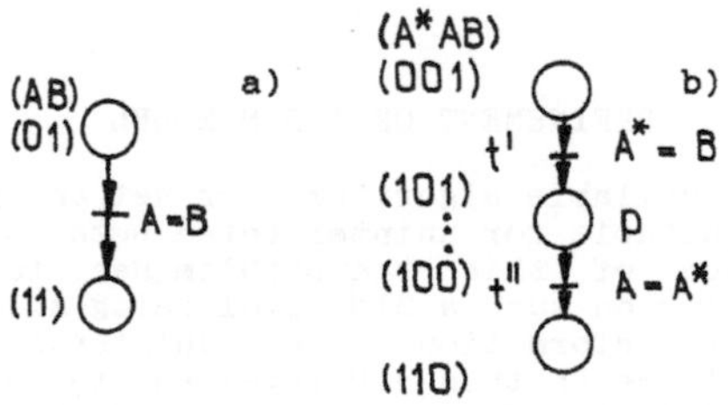

a)
(AB)   (A*AB)
(01)   (001)   b)
                        t'   A* = B
                (101)
                  :   p
A=B   (100) t"   A = A*
(11)
                (110)

Fig. 3. Modelling of an interruptable operation

The direct switching of the transition t in Fig. 3a results in a change of the value of the shared variables A and B from (0,1) to (1,1). But if the assingment is realized in the computer in two steps: testing of B and after that changing of the value of A, this behaviour has to be modeled in the Fig. 3b manner. The switching of the transitions t', t'' can result in a change of the value of A and B from (0,1) to (1,0), if the value of the shared variable B is changed after switching of t' by another parallel process.

The result of the refinement is a SIN model being on the one hand completely regarding the information about the software properties researched and on the other hand suitable for transformation into a PN.

TRANSFORMATION

In transforming a SIN into a PN both program variables inclusively their values and program statements must be replaced by net elements. Concerning the problem class discussed here a SIN can be transformed into a PN as follows:

- for each program variable v a pair of data places (p,p') has to be introduced. Thereby the value W(v) of the variable v is equal the number of tokens in p, and K-W(v) is the number of tokens in the complementary place p' respectively (K=capacity of p,p')
- each interpreted transition has to be substituted by an uninterpreted subnet with test edges and flow edges leading to data places. Test edges describe the testing of values of variables, flow edges realize the assigning of new values.

The transformation is demonstrated for dcf's and daf's in Fig. 4 and Fig. 5.

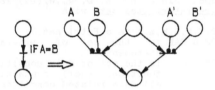

Fig. 4. Transformation of a data condition function

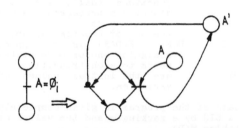

Fig. 5.   Transformation of a data action function

The above mentioned subnet substituting a dcf will be created in the following way :

- a transition must be introduced for each set of values of the dcf input variables
- test edges to corresponding data places secure, that one transition at the most gets concession for switching in a given case.

In other words:If the dcf provides "FALSE" no transition of the subnet switches.

A subnet substituting a daf can be created as described for dcf substitution. The only difference consists in additional consideration of output variables by means of flow edges. In the same way variables with discrete values (e.g. Integer variable) can be represented by a pair of data places (p,p') with K(p),K(p')>1. In a closer sense all values of variables managed by a real computer could be modelled in this way, because they are discrete ones.

A given SIN has been transformed to a PN being equivalent regarding to its behaviour.

ANALYSIS

The following sequence of activities is necessary for verification of programs using SIN model.

- refinement of the problem
- modelling of the program components
- algorithmic reduction of the model
- transformation of the verification goals on SIN- and RG- level
- transformation of the model and computation of the rechability graph
- analysis of the RG
- reverse transformation of the results of analysis on the level of program

The interpreted elements of a given SIN, i.e. daf´s and dcf´s, have to be mapped into additional net-constructs being suitable for analysis by the methods mentioned above. For most verification goals being of interest, this results in a computation of the complete reachability graph followed by an investigation of graph properties, e.g. paths through the graph or strongly connected subgraphs. If necessary, this has to be done in iteration cycles. In this sense the subjects of analysis on SIN-level must be mapped to corresponding subjects on RG-level, i.e. analysis must be done with respect to the latter. After analysing the RG the results will be retransformed onto the SIN level. Table 1 shows the connection between selected verification goals on program level and corresponding properties of the SIN or RG respectively.

Reaching a high efficiency in software verification requires a computer-aided realization of all steps from the refinement of the problem to the reverse transformation. A software tool meeting this requirement is under development.

EXAMPLE

The algorithm of DEKKER solves the problem of a correct entry of two concurrent processes into a critical program segment without using privileged operations (semaphores, monitors) but only with the help of global variables. Figure 6 shows the SIN model of the algorithm and Table 2 the transformation of the verification goals.

The proof of the correctness of the algorithm can be decomposed into 3 parts (Ben-Ari,1985):

- the correctness of the mutual exclusion
- the non-existence of deadlocks
- the non-existence of obstructions.

Already for this small problem the mathematical proof is very expensive and transparency is lost.
The computation of the reachability graph results in 122 nodes and 244 arcs. The verification goals shown in Table 2 were proved to be true by analysis of the rechability graph.

By the same way the correctness of a communication protocol for a multiprocessor operating system was proved successfully (RG: 900 nodes,1600 arcs).

CONCLUSIONS

The aim of the paper consists in demonstrating a method for modelling and verification of software for distributed computer systems on the basis of the Petri net theory. The method can also be applied to parallel systems in general.

In comparison to other methods it is dedicated mainly to a complete verification, i.e. the proof of the total absence of failures concerning the verified properties, and not for simulation and testing. Moreover, the method is of general use for the class of problems discussed not only for special problems being of "good will".

The Software Interpreted Net (SIN) defined in this paper allows a compact modelling of programs on a problem-oriented level as well as its verification on the basis of a mathematical theory. Details of components of the verification algorithms are discussed. The method was used to prove the correctness of software components for communication between microcomputers.

This method is only limited by restictions of computing time and memory space. It can be implemented on advanced computer systems of 16 bit word length or more. A software tool is under development.

REFERENCES

Arendt,F.,H.F.Heltzig and B.Kl}he(1987). Modelling and verification of real-time software using Petri nets. _Informatik.Informationen Reporte.3._ Berlin,Heft 15,72-83.
Ben-Ari,M.(1985)._Grundlagen der Parallel-Programmierung._ Hanser-Verlag,M]nchen Wien.
Kluehe,B.(1988).Untersuchungen zur Gestaltung und Verifikation von Betriebs-system-Komponenten f]r Mehrmikro-rechnersysteme.Dissertation A, Wilhelm-Pieck-Universit{t Rostock.
Quirk,W.J.(1985)._Verification and validation of real-time software._Springer-Verlag,Berlin New York Tokyo.

TABLE 1 Analogies between Software- and Net-propert.

software properties	net properties
structural properties (failures in control flow)	structural propert. insulated elements
failures in data-handling (redundant variables overflows conflicts in accessing)	reachability invariants,capacity conflicts
failures in program structure (redundant statements deadlocks contradictions in branching infinite loops)	dead transitions dead markings conflicts loops
correctness of synchronizing (exclusivness absence of deadlocks absence of obstructions)	reachability dead markings,strongly connected parts in RG
reachability of specific correct or faulty program states	reachability coverability

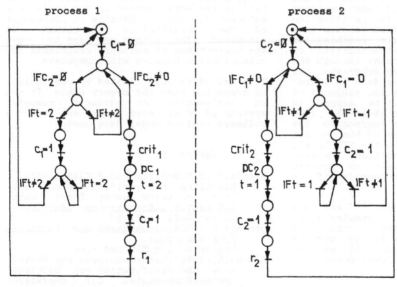

critical sections: crit1,crit2   inital values: c1,c2,t =1

Fig. 6 SIN model for the algorithm of Dekker.

TABLE 2 Verification goals for the example

program	net model
correctness of mutual exclusion of critical section crit1 and crit2	non-reachability and non-coverability of the marking $m(pc1)=m(pc2)=1$
non-existence of deadlocks: one process cannot reach its critical section	non-existence of dead markings and of loops in RG containing only crit1 or crit2
non-existence of obstructions: the two processes can reach their critical section only alternatively	non-existence of special pathes in RG

INCIDENCE OF PARAMETERS
IN THE PERFORMANCE OF A
DISTRIBUTED DATABASE

P. Blesa* and R. Zambardino**

*Depto. Sist. Inform. y Computacion, Universidad Politecnica de Valencia,
Valencia, Spain
**Department of Computing, Staffordshire Polytechnic, Stafford, UK

Abstract The performance of a real time distributed database is dicussed,
taking into account the effect of the number of copies of each data item.
A vital factor in the behaviour of these systems is the number of messages
(both control and data) produced in each case, which is often very high
and causes heavy network traffic. The paper presents abstract models for
the most important methods of concurrency control, explaining all the
assumptions made. A simulation study of the performance of these model has
been carried out, and the techniques used are explained. The results of
the simulations are plotted and compared, to show how the methods of
concurrency control, the type of environment under study, the number of
items, and the number of copies of each item affect the average waiting
time of each transaction, and the number of control and data messages.

Key words. computer simulation, concurrency control, database management
systems, distributed control, distributed databases, performance
evaluation.

INTRODUCTION

There are important reasons for the
rapid increase in importance of real time
distributed database systems. In
particular, they are capable of
eliminating many of the shortcomings of
centralized databases, and fit more
naturally in the decentralized structure
of many organizations.

There are two main contributing
factors: first, the development and
widespread use of small computers now
provides the necessary hardware support
for the development of distributed
information systems, and second, the
technology of distributed databases is
based on two other technologies wich have
solid foundations, namely computer
networks technology and database
technology.

The problem of concurrency control in
real time use of databases, in order to
preserve the integrity of the data is well
known. Literature has been published on
this subject over the last 15 years.
Simulation techniques have also been
applied to study the effects of some
parameters on the performance of Database
Systems (see [1], [10] and [11]).

With the introduction of distributed
databases, the problem of concurrency
control becomes much more complicated,
especially if the database is totally or
partially replicated. Much work has been
published in this area, and [3] gives a
good survey of concurrency control
methods.

Although distributed databases have
been studied for some time, there are
still only a few commercial systems in
existence, all having severe limitations.
However, it is our opinion that such
systems are going to be in great demand as
the necessity for using large, real time
systems increases.

Surprisingly, we have not found in
the literature any paper on a simulation
taking into account some of the parameters
that would appear to be very important in
distributed systems. The classical
problems in a centralized database (i.e.
lock granularity, level of concurrency and
types of transactions) are well studied,
but the effects of the method of
concurrency control used, and of the
number of copies of each data item in
distributed systems, are yet to be
simulated.

We are mainly interested in the
average waiting time for transactions, as
the actual use of CPU or I/O is not
usually the main problem in these systems.
Another important aspect considered is the
number of messages (both data and control)
produced in distributed systems. This is
often a critical factor in real time
systems of this type.

We also wanted to study the
performance of timestamping: this method
is seldom used, but can be very efficient
in many situations if adequate precautions
are taken.

DESCRIPTION OF THE SIMULATION

The two basic methods for preserving
the integrity of a concurrent database,
locking and timestamping, are well known.

Both of them require the database to be partitioned into items (or granules). If a locking protocol is used, transactions must take out a lock on an item before being able to read or write to it. In timestamping, transactions are given timestamps; the consistency of the database is preserved by comparing these timestamps with the read or write timestamps of the data item which are to be read or written.

In all the cases of our study we have chosen the three phase approach as a model for discrete event simulation [9], as this combines the simplicity of the activity approach with the efficient execution of the event based approach. To avoid fictitious transient conditions we allow the system to achieve a steady state, at the beginning of each simulation, before starting the accumulation of results.

It is important to state clearly the assumptions upon which our results are based. Studies published so far show seemingly contradictory results [1], due to the wide variation of the assumptions made.

Our assumptions are:

- We are studying performance from the point of view of data contention. We therefore assume that we have in all situations enough system resources (mainly CPU and I/O). The introduction of limits in system resources can change the results.

- The coordinator site sends messages to other sites in order to ask for locks, to read a data item, and so on. The duration of these messages (which is very variable in a real system) is taken at random within a specified range, with uniform distribution.

- We assume three different interactive environments, based on three specific types of transactions:
a) Read-one transactions, whose only operation is to read a data item.
b) Read-write-one transactions, wich read an item, perform some computation, and write the new value of this item.
c) 'Interchange transactions', when some quantity is interchanged from an item A to an item B. This is the classic TP1 style of transaction often used in benchmark tests.
These transactions work in the following way:
i) The coordinator site sends a message/s asking for the data item/s to read. If the operation is possible, the read value/s is/are returned (and in the case of read-one, the transaction is finished). If the operation is rejected, a message control with the negative answer is returned, and the transaction is restarted.
ii) Once the coordinator site has received the answer, it sends the new value/s to be written, waits for commit message/s, and sends the last message/s to unlock the data items.
iii) In the case of timestamping, the write operation can be rejected, and the transaction is restarted.

- We assume that longer transactions, which occur very seldom, do not affect the overall results, or are done when the system is lightly loaded. Systems where long transactions occur often, are not covered by our simulation.

- We have given different probabilities to each type of transaction, thus obtaining two types of environment:
a) Read-oriented: The system is mainly reading values from database items.
b) Write-oriented: The most frequent transactions are those that write values.

The definition of each type of environment is as follows:

Type of system	Probability of each transaction type		
	Read	Write	Interchange
Read-oriented	0.8	0.1	0.1
Write-oriented	0.2	0.5	0.3

We have used the simplest cases, i.e. Locking and Timestmping in a Centralized System, in order to introduce the basic ideas.

Locking (Centralized System)

For concurrency control with locking, the transaction starts by asking for the locks. If it is successful, read-one transactions are finished, whereas for the other two types some processing is done, and then the transaction writes and unlocks the item/s. If the lock/s is/are not granted, the transaction is rolled back and must restart later.

Transactions are assumed to follow a cyclical path as in the diagram shown in Fig. 1.
(In all diagrams [ ] is an active state, ( ) is a dead state, < > is a choice, and ---> means a flow of simulation entities).

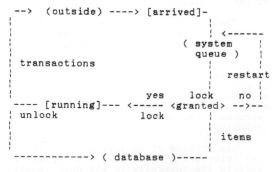

Fig. 1 Locking (centralized system)

We assume that transactions are originated from different terminals, enter the system queue, and ask for the locks they need. In this model all locks are requested at the arrival of the transaction; hence, deadlock is not possible.

If the locks are granted, the transaction begins to run, and when finished, will unlock the items. If any

lock is not granted, the transaction must
wait and repeat later the request for the
locks.

Timestamping (Centralized System)

As in the previous model,
transactions are assumed to cycle
continuously, but now the rules are quite
different (see Fig. 2).

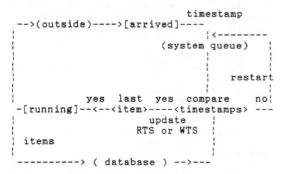

```
                              timestamp
 -->(outside)---->[arrived]----
 |                            |<--------
 |                  (system queue) |
 |                            |        |
 |                            |        |
 |                            |  restart
 |                            |        |
 |            yes last yes compare  no|
 -[running]--<--<item>----<timestamps> ---
 |                  update    |
 |                  RTS or WTS |
 | items                      |
 |                            |
 ----------> ( database ) -->---
```

Fig. 2 Timestamping (centralized system)

(RTS and WTS are the read and write
timestamp of the active data item
respectively)

Once a transaction enters the system
and receives its timestamp (ts), a
comparison must be made as follows:
- for a read operation, if ts < data item
(WTS) then the   transaction aborts
- for a write, if ts < data item (RTS),
the transaction aborts.

Some authors consider that if ts<data
item (WTS) in a write operation, then the
transaction must be aborted, but we agree
with Ullman [11], that the transaction
must not write the item, but should not be
aborted.

If all the requested reads / writes
are allowed, the transaction can run,
updating the read or write timestamps of
the items as it progresses.

Of course, transactions may be
aborted and then restarted. In these cases
we must be careful with the delay time for
restarts, because there exists a potential
risk of cyclic behaviour involving only a
few transactions. The simplest solution is
to use a random number generator to select
a random period of time during which an
aborted transaction must wait before
restarting.

LOCKING (DISTRIBUTED SYSTEMS)

In a distributed database, data items
are kept at widely dispersed locations,
and each item may have several replicated
copies.

In the use of locking for distributed
systems, different approaches have been
proposed, but only three of them appear to
have gained wider acceptance:

- Write locks all, read locks one: for a

write operation all copies of the items
involved are write-locked, whereas in a
read operation only one copy is read-
locked.

- Primary copy: each item has a designated
master copy, and all locking requests for
an item are directed to the corresponding
master copy.

- Central node: A particular node of the
database is given the responsibility for
managing all lock requests.

Other approaches for concurrency
control, such as snapshots, conservative
timestamping, classes of transactions [7],
majority locking, primary copy token [12],
have not been considered, as some of them
are not relevant, and others are
appropriate only for specific
environments.

From the point of view of our study,
the central node and the primary copy
produce similar results. The main
difference is that in the case of the
central node, most of the message traffic
is directed to or from one node. This
means messages can be bundled to and from
the central site, but of course this
method may result in the creation of a
bottleneck around the central node.

Locking with Primary Copy

In this case, the transaction
proceeds according to the following
diagram (Fig. 3):

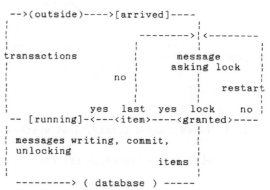

```
 -->(outside)---->[arrived]----
 |                            |
 |                -------->|<---------
 |                |          |        |
 transactions     |       message     |
 |                |       asking lock  |
 |         no     |          |         |
 |                |          |    restart
 |          yes last yes   lock   no |
 -- [running]-<---<item>----<granted>----
 |                            |
 | messages writing, commit,  |
 | unlocking                  |
 |                     items  |
 |                            |
 ---------> ( database ) -----
```

Fig. 3 Locking (distributed system)

With this approach, the locks are
sought on some node of the system by means
of messages. We have chosen the strategy
of pre-claiming all locks at the begining
of a transaction, which guarantees that
deadlocks are not possible.

Locking with write locks all, read locks one

Transactions cycle in a similar way
as in the previous model. The main
difference is that for a write operation,
a write lock is asked on each copy. All
messages are sent in parallel to avoid
long delays.

DISTRIBUTED SYSTEMS BASED ON TIMESTAMPING

The basic ideas of the timestamping approach in centralized systems have been explained in Section 2.

Most of the methods for distributed databases systems based on locking have their counterparts for timestamping. Therefore, we will study only one model: Primary copy. This strategy consists of comparing the time of the transaction with the write timestamp of the selected master copy of each item, (in the case of a read operation) or with the read timestamp (in the case of a write operation) in order to decide the fate of the transaction.

After each successful read or write, the corresponding read or write timestamp of each copy must be updated. If the transaction is aborted, the old timestamps of the items must be restored.

Timestamping with primary copy

As in the previous model, transactions are assumed to cycle as indicated in the diagram of Fig. 4.

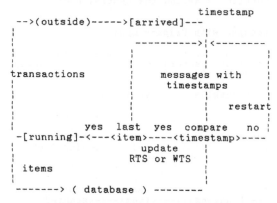

Fig. 4 Timestamping (distributed system)

RESULTS FOR DISTRIBUTED SYSTEMS

As we have stated in Section 1, we are intereted in determining the average waiting time of transactions, and the messages produced in the system.

The experiments have been carried out for a number of combinations of data items at each node, and for different number of copies of each item throughout the system.

The following cases have been investigated:

a) Four methods of concurrency control:
- Locking with primary copy
- Locking with central node
- Locking with write locks all, read locks one
- Timestamping with primary copy

b) Two types of 'environment': read oriented and write oriented.

c) Database partitioned in 100 or in 1000 data items

d) Number of copies of each data item either 4 or 20

e) Two sets of different sizes for the real-time system under consideration, defined in terms of minimum and maximum number of terminals active (i.e. generating transactions) at any time.
For a minimum of 10 active terminals, the maximum values considered were 10, 20, 40, 80 and 160.
For a minimum of 20 active terminals, the maximum values considered were 20, 40, 80, 160 and 320.
In the simulation, when any transaction finishes, the number of expected active terminals is determined by means of a random number in the range minimum-maximum, specified above. If this number is less than the transactions currently in the system, the appropiate number of new transaction arrivals is generated.

Every set of data was run several times, with different random number streams, to obtain statistically significant results. Each group of runs produced for all results their mean, standard deviation and confidence interval at 90% confidence level. The level of accuracy achieved was highly satisfactory, with intervals of the order of 0.6% of the mean.

The results obtained are plotted in the following graphs.

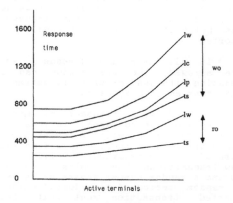

Fig. 5 Response time (100 items)

Legenda: lw =write locks all; lc =locking central node; lp =locking primary copy; ts =timestamping;
ro = read oriented systems; wo = write oriented systems.

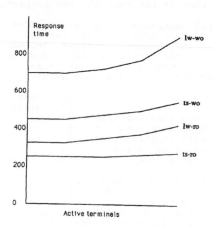

Fig. 6   Response time (1000 items)
         (for symbols see Fig. 5)

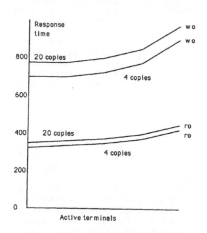

Fig. 7   Effects of the number of copies on
         the response time.
         (for symbols see Fig. 5)

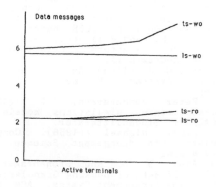

Fig. 8   Number of data messages for 1000
          items, 4 copies.
          (for symbols see Fig. 5)

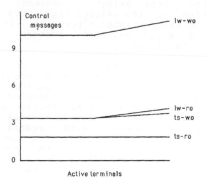

Fig. 9 Number of control messages for 1000
        items, 4 copies.
        (for symbols see Fig. 5)

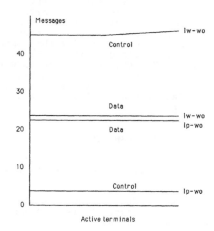

Fig.   10 Effect of the number of copies on
          the number of messages.
          (for symbols see Fig. 5)

CONCLUSIONS

Average response time

Database partitioned in 100 data items. As
can be seen in Fig. 5, the method of lock
and   write all copies is the one that
produces   the   greatest   delays   in
transaction processing, given that in
order to perform a write operation, all
copies must be locked.   In all the other
methods, only one copy is locked. The
lock-and-write-all-copies is not dependent
on a specific copy, and so has high
reliability at the expense of performance.

      More surprisingly, at first glance,
the best results in our experiments have
been attained with timestamping using a
primary copy method. However, this result
is   reasonable for short   transactions
because, with locking, once the locks have
been granted at the beginning of the
transaction, the items remain locked until
the unlock message at the end,   not
allowing any concurrency for these data
items. In the case of timestamping, the
transaction   reads   and   updates   the
timestamps of items, allowing other
younger transactions to read or write,
thus permitting higher concurrency.

For read oriented systems, the variation of the waiting time with the rate of transactions is very low. In the case of write oriented systems, the difference is higher.

Database partitioned in 1000 items. The main conclusion is that, as expected, a much higher level of concurrency is allowed in this situation, keeping the average waiting time for transaction at a reasonable level, even when transaction density is high. With regard to concurrency control methods, the comments written for 100 items can be repeated here, but as we can see in Fig. 6, in the case of read oriented systems, the response time has changed very little: with a reasonable number of items, and with many read transactions, we have very few problems of data contention.

Number of copies The only difference in the waiting time is for the method of write locks all, since in the case of primary copy and central node, a transaction must wait only for the answer of one site for each data item. As we can see in Fig. 7, the performance is worse with the increase in the number of transactions, and is even worse (obviously) in the case of write oriented systems.

Control and data messages

The results are quite different for each case, but along the lines which were expected.

Data messages The minimum number of messages for each type of transaction can be computed in this way:

a) For intechange transactions, two data messages with the contents of the two data items which have been read, and one data message for each copy of each item, with the new contents to be written.
If we call nc the number of copies, the number of data messages is then 2+2*nc, giving a value of 10 for nc = 4, and 42 for nc = 20.

b) For read-write-one transactions, one data message for the read item, and nc (one for each copy) with the new contents, i.e. nc + 1.
c) For read-one transactions, only one data message with the read value.

Of course, where the interval between two consecutives transactions is shorter, more transactions will need to be restarted, which results in more redundant data messages.

For read oriented systems, Fig. 8 shows how the effects of the rate of arrival of transactions and of the control method are not very important. For write oriented systems the situation is different: the method of central node is the best, given that most of the write messages are produced only when the central node has granted the necessary locks. On the contrary, timestamping is the worse, given that it is the only method where a transaction can be rejected

even when it has sent the new contents.

Control messages This case differs from that of data messages, in that the minimum number of control messages depends not only on the number of copies and on the type of transaction, but also on the concurrency control method used:

	Inter-change	Read-write	Read-one
Lock (and timestamping) with primary copy:	6	3	1
Lock with central node:	7	4	2
Lock write all copies, read locks one:	2+4*nc	1+2*nc	1

The actual number of control messages increases as more transactions are running simultaneously.
Figure 9 reflects the earlier comments. The case of write locks all is at the top, as we could expect.

Number of copies. In order to show how the number of copies affects the number of control and data messages produced, we have drawn in Fig. 10 a comparison between two representative methods, lock with primary copy and lock with write locks all, for 20 copies of each data item. For read oriented systems the differences are lower, but for write oriented systems, as we expected, control mesages produced for write locks all are far away in the top.

REFERENCES

[1] Agrawal, R., Carey, M. and Livny, M. (1987). Concurrency Control Performance Modelling: Alternatives and Implications. ACM TODS, Dec.
[2] Benigni, D. R.(Ed.) (1984). A guide to Performance Evaluation of Database Systems U. S. Department of Commerce, Dec.
[3] Bernstein, P. A. and Goodman, N. (1979). Approaches to concurrency control in distributed data base systems Proc. AFIPS, NCC Vol. 48
[4] Bernstein, P. A., Hadzilacos, V. and Goodman, N. (1987). Concurrency control and recovery in Database Systems. Addison-Wesley.
[5] Bernstein, P. A., Shipman D. W. and Rothnie, J. B. (1980). Concurrency Control in a System for Distributed Databases (SDD-1) ACM TODS, March
[6] Ceri, S. and Pelagatti, G. (1988). Distributed Databases. Principles and Systems. McGraw-Hill.
[7] Date, C. J. (1983). An Introduction to Database Systems. Volume II. Addison-Wesley.
[8] Peter Jennergreen, L. (1984). Discrete-events simulations models in PASCAL/MT+. Studenlitteratur.
[9] Pidd, Michael (1986). Computer Simulation in Management Science. John Wiley & Sons.
[10] Ries, D. R. and Stonebraker, M. (1977). Effects of Locking Granularity in a Database Management System. ACM TODS, Sept.
[11] Ries, D. R. and Stonebraker, M. (1977). Locking Granularity Revisited. ACM TODS, June
[12] Ullman, Jeffrey D. (1983). Principles of Database Systems. Pitman.

USING CONVERSATIONS TO IMPLEMENT RESILIENT OBJECTS IN DISTRIBUTED SYSTEMS

J. A. Cerrada, M. Collado, R. Morales and J. J. Moreno

*Departamento de Lenguajes y Sistemas Informáticos e Ingeniería de Software,
Facultad de Informática, Universidad Politécnica de Madrid,
Campus de Montegancedo, Boadilla del Monte, 28660 Madrid, Spain*

Abstract. This paper introduces an implementation of resilient objects based solely on the conversation mechanism, in the framework of the CSP scheme. Every transaction over the object is defined as a conversation involving one or more external processes and one internal process that monitors the state of the object. The proposed implementation facilitates the distribution of the transaction code over a computer network, and also an automatic control of concurrency by enforcing an adequate "readers and writers" policy. Transactions are invoqued by procedure calls, but only message exchanges are actually done between different processors.

Keywords. Distributed programming; CSP; Distributed databases; Conversations; Fault-tolerance; Resilient objects.

INTRODUCTION

In the last years, the interest for reliable and safe software systems has greatly increased. In the first approaches, the fault tolerance concept was associated only to hardware failures. Nevertheless, they are not the only kind of failures that can appear in a computing system; software errors must also be taken into account. As programs become larger and more complex, the detection and correction of software failures gets more and more important. Moreover, hardware failures and software failures can be treated with similar detection and recovery mechanisms.

The use of fault tolerance mechanisms in a program does not ensure the recovery of all the failures which may occur. Nevertheless, fault tolerance techniques can decrease or, sometimes, eliminate the effect of an error. Of course, they increase the complexity of the program, but this is the price to be paid for a realiable application.

Probably, fault tolerance in distributed systems and real-time applications is the most interesting topic in this field, and many works are devoted to them. The approach taken in this paper is to build a concurrent system as a transaction-based system (Bernstein, Hadzilacos and Goodman 1987). A transaction is a basic change of the permanent state of the system.

A well known proposal for handling failures is based on atomic actions and atomic objects (Liskov and Scheifler, 1983, Reed 1983). An atomic action must appear as indivisible to the external world. Viewed as a transaction, it must either succeed doing its desired effect, or abort without any effect. A kind of atomic action that involves several processes is a conversation (Randell, 1975). A more structured approach to conversations, based on Hoare's CSP scheme (1978) introduces conversation modules (Collado, Morales and Moreno 1989).

An atomic object is a data abstraction operated only by atomic actions. Because an atomic object can recover its previous state after a failure, it is called also resilient object.

Conversation modules can be used as a foundation for the implementation of atomic objects. The relationship between conversations and atomic objects in distributed programming is the same as between functional abstractions and abstract data types in sequential programming. From this point of view, operations on resilient objects can be defined as conversation modules. The idea is to model the object as an entity monitored by an internal process, which communicates with the external processes that request services from the object.

Table 1 summarizes the special versions of some basic abstraction mechanisms that are needed in fault tolerant distributed programming.

TABLE 1. Abstraction mechanisms (and implementations).

Sequential programs	Distributed programs
Functions (FUNCTION, PROCEDURE)	Actions (Atomic actions - Liskov, FT-Actions - Jalote, Convers. mod.- Collado)
Abstr. data types (PACKAGE - Ada, CLASS - Simula, MODULE - Modula-2, class - C++)	Atomic objects (Guardians - Liskov, Resil. obj.- this paper)
Programs (PROGRAM)	Processes (PROCESS, TASK)

These mechanisms are explained in the following sections.

CONVERSATIONS, FT-ACTIONS AND CONVERSATION MODULES.

Following Randell's original proposal in (1975), a conversation is defined as a distributed control structure in which a set of processes can participate in a coordinate way. Processes can communicate between themselves inside a conversation, but cannot communicate with any process outside the conversation.

A conversation defines an atomic action in the sense of (Liskov and Scheifler 1983), or a transaction in the sense of (Berstein, Hadzilacos and Goodman 1987). It ensures that the common action is executed or, if it is not possible, the state of the processes is not changed at all.

FT-Actions (Fault Tolerant Atomic Actions) are a special kind of conversations introduced in (Jalote and Campbell, 1986) and based on Hoare's CSP scheme (1978). As in conversations, the communication between a process inside a FT-Action and a process outside this FT-Action is forbidden. The mechanism of FT-Actions allows forward as well as backward error recovery. Some additional features of FT-Actions are:

- Atomicity.
- Recovery Line for Backward Error Recovery.
- A Test Line for the Processes.
- Recovery Measures.
- Nesting of FT-Actions.

In order to use a good programming methodolgy, some features of FT-Actions must be improved. In particular, the distribution of code for the actions over the processes, and the fixed binding of process names to conversations. To avoid this, we have presented Conversation Modules (Collado, Morales and Moreno, 1989), a new approach based on treating conversations by means of FT-Actions but encapsulating their code.

Conversation modules are independent packages that encapsulate the code for a FT-Action. A conversation module defines a series of members (those that participate in the common FT-Action) and contains the code for each one. The access to a conversation is obtained by calling one of its member procedures.

Using an Ada-like notation, conversation modules can be described by separating the interface part and the implementation part.

The interface part includes the specification of the conversation. It contains the headers of the members and the exported exceptions as follows:

```
CONVERSATION Conversation_Name IS
   << Exported exception names>>: EXCEPTION;
   MEMBER Member_1 (Parameters);
      ...
   MEMBER Member_N (Parameters);
END Conversation_Name;
```

The implementation part must describe the code for the alternatives of each member in the following way.

```
CONVERSATION BODY Conversation_Name IS
   ...
   MEMBER Member_i (Parameters) IS
   ...
   BEGIN
      // save state //
      ENSURE << acceptance test >>
        BY << first alternative >>
          { << Handle local exceptions >> }
        ELSE BY // restore state //
           << second alternative >>
        ....
        ELSE // restore state //;
                RAISE << exported exception >>;
      END ENSURE;
   END Member_i;
   ...
BEGIN
   << Initialization statements >>
END Conversation_Name;
```

The code between <<...>> is application dependent, and must be written by the user. The code shown as //...// is for recovery purposes, and should be generated by the compiler.

A member of the conversation is invoked by importing the conversation (by means of a WITH statement) and issuing a call:

ConversationName.MemberName (parameters)

When all the members finish an alternative in a conversation, an acceptance test is checked-out. The purpose of this test is to ensure the synchronization of the exits of all the members inside the action as follows: either all the processes inside the action end correctly or, if one of them notices an error, all of them remain in the conversation trying another alternative. When all the alternatives fail, an exception is raised in order to handle this error externally. Following this idea, the conversation ensures that all the members end with the same exception raised if the action is not possible. Of course, every member in a conversation must have the same number of alternatives to recover errors.

The implementation of a conversation could handle a failure as soon as possible. Hence, when a member of a conversation notices an error it can notify it to the rest of the processes immediately. The processes can quickly abandon the current alternative and either handle the error or try another alternative.

A summary of the remaining main features of conversation modules follows. Some of them are inherited from FT-Actions

1.- Each conversation defines a FT-Action. Hence, it has the same good properties.

2.- Conversations modules encapsulates all the code for the action they perform.

3.- Conversation members implement abstract roles, not tied to specific processes when the conversation is defined.

4.- Conversation modules can be viewed as functional abstractions, and so they can be good candidates for reusing.

5.- The conversation begins execution when all its members have been invoked.

6.- Conversations modules can be nested, avoiding error propagation. Each conversation has its own copy of the state. Only the outermost conversation can change the permanent state of the system.

7.- Only processes participating in an external conversation can participate in an internal nested conversation.

8.- Conversation modules are implicitly <u>reentrant</u> as they effectively implement functional abstractions, in the sense that they lack of permanent state (permanent variables are always passed as parameters). If a conversation is in progress and some process invokes it, a new instance of the conversation is generated.

9.- As a practical point, an implementation could detect and recover from a failure caused by a <u>deserter process</u>.

Inside a CSP framework, interprocess communication is achieved only via message exchanges. Although conversation members are called as procedures, they can only cooperate by exchanging messages through channels defined inside the conversation module. These channels are the only global items visible to all the members; shared variables are explicitly forbidden.

A message exchange is carried out by send and receive primitives:

  send(channel, message, success)
  receive(channel, destination_var, success)

The third parameter is used to notify either the success or the failure of the operation.

ATOMIC OBJECTS

From the methodological point of view, when fault tolerance is needed the concept of data abstraction leads to the idea of atomic object. For recovery purposes, shared objects in distributed systems should take the form of atomic objects.

An atomic object is an abstract data object whose operations are atomic and hence indivisible. If the object is shared among several processes, this implies that operations on the object must have the <u>serializability</u> property, i.e., the final state of the object after a series of operations must be reachable by a sequence of non-overlapping atomic operations. From this point of view, an atomic object is like a <u>monitor</u> (Hoare, 1974). Every atomic operation must either <u>commit</u> or <u>abort</u>. In the first case the job is done and the permanent state of the object is updated; in the second case the state is not changed at all. An object which recovers its original state after a failure is called a <u>resilient object</u>. In the literacy, several conceptual models of atomic objects have been proposed; see, for instance, (Liskov and Scheifler, 1983, Yonezawa and Tokoro, 1987).

One of the most important topics is that of nested atomic objects. When atomic operations are invoked, the nested (inner) object must remain locked until the outermost action ends. This is because a failure in the nesting (outer) levels must be recovered by restoring the initial state, thus aborting nested operations previously commited (Berstein, Hadzilacos and Goodman, 1987).

A classical and simple way to implement nested atomic objects is by using monitors that define the locking primitives <u>Lock</u>, <u>Commit</u> and <u>Abort</u> on every object. Any use of an atomic object must first lock the object. At the end of the global operation, either the commit or the abort procedure must be invoked, unlocking the object at this time.

IMPLEMENTATION OF RESILIENT OBJECTS BY USING CONVERSATION MODULES

The previous section has established that the operations on atomic objects must be atomic actions. Moreover, in sequential programming, an abstract data type is modeled by the definition of the type and a collection of functional abstractions. Both ways yields to a similar idea for distributed programming: the use of conversation modules in the implementation of resilient objects.

More exactly, we propose the existence of a (unique) process in each object responsible for the state of the object. This process is called the <u>demon</u>. Each operation on a resilient object takes the form of a conversation module with some members allowed to be called externally and a distinguished member assumed by the demon. These conversation modules are slightly different from which we have defined previously; they will be called <u>transactions</u> in the sequel.

Following this idea, the special member of a transaction is the only member with access capability to the internal state of the object and its execution begins as soon as the transaction is activated. This member is called the <u>local member</u> of the transaction.

The use of such these transactions is as follows. The normal (external) members can perform any operation except for using the state of the object. Any access to the object state must be done by exchanging messages with the local member. At any time, the local member can modify the internal state of the object. Because this local member has, inside the transaction, the same status as the other members, it can notice and notify any error and start the recovery mechanism of the transaction (conversation).

Figure 1 shows the internal structure of a sample resilient object with two defined transactions (one of them with two external members). As will be explained later, the code can be distributed over a set of processors by cutting this structure along lines that cross only message exchange links.

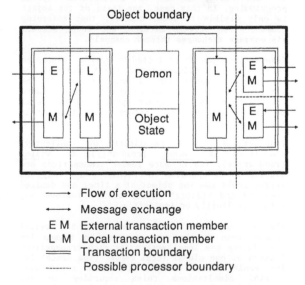

Figure 1. A resilient object structure.

Using an Ada-like notation again, a resilient
object can be described as:

Interface definition

```
RESILIENT OBJECT TYPE object_type_name IS

   TRANSACTION  transaction_1 IS
     <<Exported exception names>>: EXCEPTION;
     MEMBER  t1_member1( parameters );
     MEMBER  t1_member2( parameters );
     . . . . . . . .
     MEMBER  t1_memberN( parameters );
   END transaction_1;
     . . . . . . . .
   TRANSACTION  transaction_M IS
     <<Exported exception names>>: EXCEPTION;
     MEMBER  tM_member1( parameters );
     . . . . . . . .
   END transaction_M;

END object_type_name;
```

Object instance declaration

```
Object_Instance: object_type_name;
```

Object implementation

```
RESILIENT OBJECT BODY object_type_name IS
   << object state declaration >>

   TRANSACTION transaction_1 IS
     << inter-member channel declarations >>
     MEMBER  t1_member1( parameters ) IS
       << code for this member >>
     END t1_member1;
     . . . . . . . .
     LOCAL MEMBER IS
       { WHEN << transaction_1 guard >> => }
         << code for local member >>
     END LOCAL;
   END transaction_1;
     . . . . . . . .
   BEGIN
     << object state initialization >>
   END  object_type_name;
```

Usual visibility rules of Ada are modified in order
to use this schema in distributed system
programming. In this case, the state of the object
is only visible to local members, thus enforcing
communication between external and local members
via message exchanges through channels.

Transactions can be protected by guards, expressed
as conditions involving the internal state of the
object. The transaction is only started if the
guard is satisfied. According to the proposed
visibility rules, the guard appears in the local
member, although it refers to the whole
transaction.

Transactions can be (explicitly or implicitly)
classified in reader (read-only) or writer
(read-write). Two or more reader transactions are
allowed to be executed concurrently, but only one
writer could use the object at a time. Any desired
"readers and writers" policy can be implemented,
either explicitly or implicitly.

The possible simultaneous execution of several
transactions at the same time was a reason for
introducing one local member for each transaction,
instead of one global local member (the demon) for
the whole object. Hence, the demon could play
several simultaneous roles according to the
transactions in execution. Figure 2 is a schema of
a general concurrent transactions implementation.

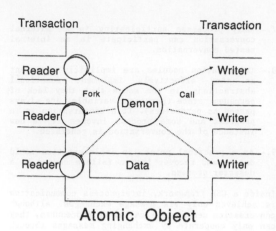

Atomic Object

Figure 2. Implementation of concurrent
transactions.

The demon is a process that waits for transaction
calls. The code for this process (figure 3) can be
a non deterministic selective reception (a guarded
command in the sense of Dijkstra and Hoare, or a
SELECT statement in Ada). If the requested
transaction is a reader one, the demon creates (by
using a fork primitive) a process that takes the
role of the local member of the transaction. If the
requested transaction is a writer one, the demon
itself takes the role of the corresponding local
member, after waiting for termination of any active
reader transaction.

```
LOOP
  SELECT
    WHEN << transaction_1 guard >> =>
      ACCEPT  transaction_1 DO
        CASE  << transaction_1 type >>  IS
          WHEN  // reader //  =>
            Start_Process( Transaction_1.LOCAL )
          WHEN  // writer //  =>
            << wait for readers finish >>;
            transaction_1.LOCAL;
        END CASE;
      END;
    . . . . .

  END SELECT;
END LOOP;
```

Figure 3. A code schema for the demon process

The code of the external members can be distributed
and/or replicated over different processors. A high
degree of parallelism can then be achieved by
executing the code of the external member in the
remote (calling) processor. This parallelism is
explicit and hence controllable by the programmer.
Also, large data structures can be passed as
parameters to external members without a high
transmission cost because in fact the call to the
external member is made locally, and only the
neccesary data parts could be passed in
inter-member messages.

Notice that this implementation of a resilient
object is active (a process guards the data) while
classical object implementations are usually
passive or static. The active model has some
advantages in distributed programming. The most
important one is that it ensures an adequate
recovery support in the case of remote node
failures, beacause the demon remains active. In
addition, an object could choose any free processor

for its execution, instead of having an unique home processor.

EXAMPLES

The first simple example models a fund transfer between bank accounts. Previous versions of this example can be found in (Reed, 1983, Campbell and Randell, 1986). The resilient object is the Bank_System, which controls a set of bank accounts. Two transactions are used in this example: **Credit**, that deposits a certain amount of money into an account and **Transfer** that makes a fund transfer between two different bank accounts. The interface part is:

```
RESILIENT OBJECT TYPE Bank_System IS
  TRANSACTION  Credit  IS
    Error: EXCEPTION;
    MEMBER Make( acc_number: Account;
                 amount: Money );
  END Credit;

  TRANSACTION  Transfer  IS
    Error: EXCEPTION;
    MEMBER Make( from_acc: Account;
                 amount: Money; to_acc: Account;
                 dest: Bank_System );
  END Transfer;
END Bank_System;
```

The Credit transaction accepts an order to deposit some money into an account. The local member updates the bank account when the Make member notifies the amount and the account number.

The Transfer transaction is more complicate. The Make member notifies to the local member which account has to be charged, and then orders a Credit transaction on the other bank. The involved accounts can be updated in parallel. If something goes wrong in these operations both accounts remain unchanged (due to the transaction nesting).

```
RESILIENT OBJECT BODY Bank_System IS
  << Account state declarations >>

  TRANSACTION Credit IS
    loc_ch: Channel;
    MEMBER Make (acc_number: Account;
                 amount: Money) IS
      done: Boolean;
    BEGIN
      ENSURE done
        BY Send(loc_ch, (acc_number,amount), done);
        ELSE RAISE Error;
    END Make;


    LOCAL MEMBER IS
      acc_number: Account;
      amount: Money;
      done: Boolean;
    BEGIN
      ENSURE done
        BY Receive( loc_ch, (acc_number,amount),
                    done );
          << Increase "acc_number" balance by
             "amount". Set "done" to false if not
             possible >>
        ELSE RAISE Error;
    END LOCAL;
  END Credit;
```

```
TRANSACTION Transfer IS
  loc_ch: Channel;

  MEMBER Make( from_acc: Account;
               amount: Money; to_acc: Account;
               dest: Bank_System ) IS
    done: Boolean;
  BEGIN
    ENSURE done
      BY Send (loc_ch, (from_acc,amount), done);
         dest.Credit.Make( to_acc, amount );
      EXCEPTION
        WHEN dest.Credit.Error => done:=false;
      ELSE RAISE Error;
  END Make;
  LOCAL MEMBER IS
    acc_number: Account;
    amount: Money;
    done: Boolean;
  BEGIN
    ENSURE done
      BY Receive( loc_ch, (acc_number, amount),
                  done );
        << Decrease "acc_number" balance by
           "amount". Set "done" to false if not
           possible >>
      ELSE RAISE Error;
  END LOCAL;
END Transfer;

BEGIN
  << Initialization >>
END  Bank_System;
```

The process that wants to make a money transfer from Source_Bank to Dest_Bank simply executes:

```
Source_Bank.Transfer.Make( from_acc, amount,
                           to_acc, Dest_Bank );
```

The second example is a classical instance of the readers and writers problem: an airline reservation system.

The system contains information about the flights of a single airline. Initially, each flight has a number of seats available. The system can accept two kind of commands: **Info**, to find how many seats are available, and **Reserve** to make a reservation of one seat if possible.

The system is implemented as an object with two transactions:

```
RESILIENT OBJECT TYPE Airline IS
  TRANSACTION  Info  IS
    Error: EXCEPTION;
    MEMBER Get( flight_number: Flight;
                number_of_seats: OUT Seats );
  END Info;


  TRANSACTION  Reserve  IS
    Error, No_Seats: EXCEPTION;
    MEMBER Make( fligth_number: Flight );
  END  Reserve;
END Airline;


RESILIENT OBJECT BODY Airline IS
  << Flight seats declarations >>

  TRANSACTION Info // reader // IS
    loc_ch1, loc_ch2: Channel;
    MEMBER Get( flight_number: Flight;
                number_of_seats: OUT Seats );
      done: Boolean;
```

```
BEGIN
   ENSURE done
      BY Send( loc_ch1, flight_number, done );
         Receive( loc_ch2,
                  number_of_seats, done );
      ELSE RAISE Error;
END Get;

LOCAL MEMBER IS
   flight_number: Flight;
   number_of_seats: Seats;
   done: Boolean;
BEGIN
   ENSURE done
      BY Receive( loc_ch1, flight_number, done );
         << Returns the number of available
            seats in "number_of_seats" >>
         Send( loc_ch2, number_of_seats, done );
      ELSE RAISE Error;
   END LOCAL;
END Info;


TRANSACTION Reserve // writer // IS
   loc_ch1, loc_ch2: Channel;

   MEMBER Make( flight_number: Flight );
      done, free: Boolean;
   BEGIN
      ENSURE done
         BY Send( loc_ch1, flight_number, done );
            Receive( loc_ch2, free, done );
            IF NOT free THEN RAISE No_Seats;
         ELSE RAISE Error;
   END Make;

   LOCAL MEMBER IS
      flight_number: Flight;
      done: Boolean;
   BEGIN
      ENSURE done
         BY Receive (loc_ch1, flight_number, done);
            << Set "free" according to seat
               availability >>
            Send( loc_ch2, free, done );
            << Reserve a seat, if possible >>
         ELSE RAISE Error;
   END LOCAL;
   END Reserve;

BEGIN
   << Initialization >>
END Airline;
```

Info and Reserve requests to the airline **Airln** take the form **Airln.Info.Get(flight, n)** and **Airln.Reserve.Make(flight)** respectively.

Transaction Info is a reader transaction, so several of them can be executed concurrently, while Reserve is a writer transaction and only one can be executed at a time.

The problem can be generalized to a series of different airlines, which could be considered as a distributed database.

CONCLUSIONS

Resilient objects are a very adequate abstraction mechanism for distributed systems programming. In this way, fault tolerance support is added to other benefits of object oriented development.

Although the communication between members is programmed via low-level message exchanges, it is done only inside the encapsulated code. Externally, service requests appear as (remote) procedure calls.

A high degree of parallelism can be achieved by distributing the code over a set of processors. At the same time, large data structures can be passed as parameters to external members without a high transmission cost.

The possibility of the classification of the transactions in reader or writer, and the implementation of a internal concurrent "readers and writers" policy increase the degree of parallelism in a way transparent to the user.

The proposed active implementation of resilient objects has clear advantages in order to recover from node crashes and also to balance load over processors.

REFERENCES

Berstein, Ph.A., V. Hadzilacos, N. Goodman (1987). Concurrency control and recovery in database systems. Addison-Wesley.

Campbell, R.H., Randell, B. (1986). Error recovery in asynchronous systems. IEEE Trans. Softw. Eng., vol. SE-12, no. 8, pp 811-826.

Collado, M., R. Morales, J.J. Moreno (1989). Conversation modules: a methodological approach to the design of real time applications. (To be presented in) Euromicro 89, Cologne, Sep. 4-8.

Hoare, C.A.R. (1974). Monitors: an operating system structuring concept. Comm. of the ACM, vol. 17, no. 10, pp 549-557.

Hoare, C.A.R. (1978). Communicating sequential processes. Comm. of the ACM, vol. 21, no. 8, pp 666-677.

Jalote, P., R.H. Campbell (1986). Atomics actions for fault-tolerance using CSP. IEEE Trans. Softw. Eng., vol. SE-12, no. 1, pp 59-68.

Liskov, B., R. Scheifler (1983). Guardians and actions: linguistic support for robust, distributed programs. ACM TOPLAS, vol 5, no. 3, pp 381-404.

Randell, B. (1975). System structure for software fault tolerance. IEEE Trans. Softw. Eng., vol SE-1, no. 2, pp 220-232.

Reed, D.P. (1983). Implementing atomic actions on decentralized data. ACM Trans. Comp. Sys., vol. 1, no. 1, pp 3-23.

Yonezawa, A., M. Tokoro (Ed.) (1987). Object-Oriented Concurrent Programming. Addison-Wesley.

A DESIGN MODEL OF DISTRIBUTED
DATABASE IN REAL-TIME CONTROL

Fang Bin and Peng Tian Qiang

Automation Research Institute of Metallurgical Ministry, Beijing, PRC

Abstract. The problem of traditional distributed database implementation in real-time
control is discussed. According to features of data in real-time control distributed
system, a enhanced performance design model (EPADM) of distributed database control
is presented. Lastly, measurement analysis and some appraizement on EPADM is gived.

Keywords. Distributed control system; distributed database: real-time control;
data distributity; communication network; enhanced performance design.

INTRODUCTION

In recent year, distributed computer control systems are
extensively developed in real-time control. In distribute
control system, the control functions are distributed
among many local control units connected by the communication network, and go concurrently. For processing
efficiency reason, the data should also distributed
among local control units. The closer the data to the
processing, the higher the efficiency is. In other
hand, many data in distributed control should be shared
by many local control units. Processing in one cite may
be interested in data in other cite. The solution to
meet the two requirements is to develop distributed
database. The distributed database provides a global
database logically, and can be accessed by each local
control unit, so as to achieve data sharing objective.
Physically the data are spread over the local control
units, so that the data are close to the processing,
achieving the processing efficiency objective.

We have many studies in distributed database in information processing. According the ISO OSI reference model,
distributed database is implemented in application layer
, and data often stored in secondary storage(whenchester
disk, floppy disk, etc). In traditional design, the data
query process is : first check if the data is in the
cite, if not , then fetch it from other cite data resides
through network, then return back to user application(
refer to Fig.1.)

When we use traditional method design distributed database of real-time control system, the first problem faced
is the long access response time can hardly applied to
real-time control. The host of local unit should take
enormous time to process distributed transaction. The
second problem is the complicated design system can not
implemented in local controller, for many local control
units are lower end processer, and have limited memory
and processing capacity. Many secondary storages can
hardly suite for harsh industrial environments. Therefore
, we should find a simple and efficient design method
used in real-time control distributed system.

In my opinion, the main objective of the design of a
distributed data base in real-time control is the
efficiency. Therefore, we should place data to its
processing as close as possible, achieving high processing locality. Also, the system should have high reliability and high availability. Secondly, comparing to the
information processing, the data quantity of real-time
control are much less, the key is efficient handling of
the limited information. Based on above idea, I would like
to present a enhanced performance architecture design
model (EPAPM) of real-time distributed database based
on industrial local area network and do some analizing
and appraising on it.

REQUIREMENT IN HARDWARE AND NETWORK

let's consider distributed database with computer communication network. Fig.2. is a implementation model of
a distributed system. It can be divided into two subnet-

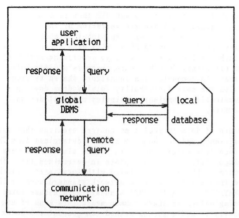

Fig.1.Traditional query process

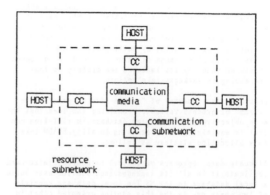

Fig.2. A model of distributed control system

work ; communication subnetwork and resource subnetwork. Communication subnetwork is responsible for information exchange between computer hosts. CC is communication controller(called IMP in ARPA network). Resource subnetwork consists of all the host computers (or controllers). Traditional distributed database is implemented in resource subnetwork , the communication subnetwork only used to exchange information.

We define a concept, "data distributity", that is access a data need to do distribute processing. Data has no distributity, that is the data is just in the own cite, user can access it like access a data in centralized database. In traditional distributed database implementation, data distributity is in resource subnetwork, so the host is always perplexed by the distributed transaction, and high efficiency can hardly achieved. Drop the data distributity down to communication subnetwork, make the host free from distribute transaction, is the start point of design model present here. If so, the data seen by the host are no distributity, the host access the data is like access a individual centralized database. The distributed processing is done by communication subnetwork. To implement these, we should have a database, in one hand, is accessed by host; in other hand , it is used to do distribute processing by communication network. The storage of database should be shared by host and CC. Thus, we have first requirement of design, the hardware requirement: building a shared storage(SS) to store the database. The host and CC can access SS independently. SS can be battery backup RAM, bubble memory, and the like.

The second requirement of the design model is that the network can do broadcasting service, that is a data frame can be transmitted onto network and received by all the stations. Modern local area network, especially that used in industrial control, can provide this service. IEC PROWAY standard provide GSD broadcasting service. New developed MAP network provide broadcasting service by broadcasting address (or group address).

These two requirements above is the base of the design model EPADM.

DATA ALLOCATE STRATEGY

The first problem faced by design distributed database is data allocation. First of all, let's study the features of data in real-time control, that is features of process variables in distributed control system. Typically, the variable is produced in some stations of the system (collected from industrial process, or calculated by the processing, or set by human, etc), these stations are the sources of the variable. The variable is used by other stations of the system, these stations are the leakages of the variable. As a matter of fact, a variable usually has only one source (leakage may be more than one), that is process variables have a feature of single source. The EPADM assume the data all are single-sourced. Yet, multi-sourced data can be defined as multiple single-sourced data. The process variables also have a stable stream feature. The process variables usually have no existance-variaton, they only have value-variation. They exist from begin to end, but their value periodiclly variate. The data sharing by multi-stations is value sharing. The change of value of the data in source should be known by its leakages immediately, so that processing in leakages take actions.

Considing the features of "single source" and "stable stream" of process variables, considing the idea that the main objective of distributed database in real-time control is maximizing the processing locality, EPADM take data allocation strategy as follows:

Allocate data (process variables) in source station, and replicate it in all its leakages. Each station user build a local database in its SS. The local database consists of two parts. one is the this station oriented (that is source is this station), called TOD. Another part is other

stations oriented (this station is the leakage) data, called OTD. Thus, all the data used by this station are in this station own storage SS, data is close to the processing, achieving a totally processing locality.

DATA CONSISTANCY UPDATE AND DATA ACCESS

As stated above, EPADM replicate the data in all stations using it, to achieve a totally processing locality. However, how to gurantee the data consistancy among so many redundancy backups. In EPADM, this is done by communication subnetwork.

We have assumed that the communication network provides data broadcasting service. From the point of effectiveness, updating the redundancy backups by data broadcasting is the best method. EPADM select it. Each station CC periodically broadcasts the data of TOD in its SS, other stations if interesting in the data, receive the data; and update its OTD. These consistancy updating manage mechanism go forever, and is completely independent to the host data access. We known that broadcasting service of network has no guarantee of data reception. Usually , point-to-point data transfering assume positive acknowlegement with retransmission (PAR) technology, that is transmitter not receive acknowlegement from receiver until time out, it retransmit the original frame. These process repeats until data is correctly received by destination station or multi-retransmission failed and abort. The data updating in EPADM is periodically going. We can think it as a spontaneous retransmission. This spontaneous retransmission is a compensation of the no guarantee data transfering of broadcasting service.

Through this kind of data broadcasting update, each local database combined into a global database (Fig.3).

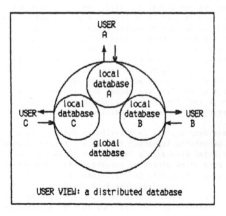

Fig.3. Local database and global database

Through defination of TOD and OTD, that is a set of global database and dynamic variation of the set, one station can access any data in global database, thus achieveing data sharing. Physically data decentralized to each station, data is close to processing, achieving high efficiency. In other word, there is a distributed database in communication subnetwork shared by each station. The data distributity totally drop down to communication subnetwork, achieving original design objevtive.

Feature of data in real-time control requires that data consistancy update should guarantee determined time. That is once data value changed in source station, the data in leakage stations should update in determined time. This problem of EPADM is associated with local area network used. Industrial local area network standard, such as IEC PROWAY, IEEE 802.4 that MAP standard used, assume token passing method to share common media. This kind of non-competitive media access method can guarantee a determined maximum response time after a data transmission

request. In fact, it is the time required that token round
the network a circle. EPADM data broadcasting update is
done by the network data transmission, so the maximum
updating time is therefore determined. Once the variable
in source station being changed, broadcasting update
mechanism produce a broadcasting request immediately,
after a determined time, the variable in all its leakage stations are updated. May be the variable is broadcasted every two token round circles(depending on the
quantity of TOD in source station, and priorites requirement), but the updating time is determined.

Another feature of real-time control data is its various
time requirements. Some data should be updated in milliseconds, some only require not more than one minute. For
this, EPADM take a priority strategy. Giving a priority to
each variable (set in creating definition). The higher
the priority, the higher the updating frequency. In detail
, that is the highest priority data updates in each token
pass circle, the lower priority variable is updated in
multiple token circle time(a counter is enough to implement this). This simple priority design strategy can make
use of network bandwidth more rational.

For saving the network bandwidth, EPADM divide one data
into two parts, one is its update part, which should be
updated continously; another part is description part,
which is set when data being defined, and can hardly
change, such as a varible description message, etc. Only
the update part of data is broadcasted on network. In
another hand, the data update can also be designed that
it is to be broadcasted determined times after its
value being changed in source station, and then stop
its broadcasting until its next value change. Many strategy can be used to optimize the design.

Last, let's see the data access. For the data allocation
strategy and data consistancy updating stated above, the
host access the database become very easy. It is the host
access its own storage SS, and response can be almost immediately, this brings high host processing efficiency.
Another data updating mechenism executed by communication controller proceeds total independently(Fig.4).
All the query optimization technology and complex
accessing structure can be used here(may be not nessary)
. No problem of distributed transaction, no problem of
distributed concurrency control, etc. All is just like a
centralized database.

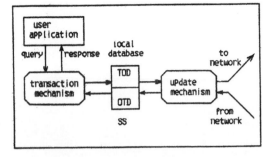

Fig.4. Data access and update

MEASUREMENT ANALYSIS

I have present a real-time control distributed database
desine model EPADM. It is based on storage sharing and
data broadcasting. EPADM provides a distributed database
that is shared by every station hosts of network. Physically, data is stored in each station storage SS, providing a total processing locality.

There are two new questions arised:

.if the frequency of data broadcasting update is
acceptable by real-time control distributed system.
.if the capacity of database is acceptable by
real-time control distributed system.

To answer these questions, I am going to do some measurement analysis on EPADM.

In EPADM, data is replicated in each station use it. The
data redundancy consistancy update is done by broadcat-
ing provided by network service. To do analysis, we consider a EPADM implementation based on IEC PROWAY standard
network. IEC PROWAY standard assume token passing method
to share common media(refer to IEC PROWAY standard). The
token is a broadcasting frame. Each time a station hold
the token, it can transmit a point-to-point frame, and
broadcast a frame (as data part of token frame). That, the
broadcasting banbwidth is about a half of the total
bandwidth. The data rate is 1 Mbps. PROWAY token pass
strategy can give 86% utilization. That is its data
throughput can achieve 86K bytes/sec., and data broadcasting throughput is about 43K bytes/sec.

In real-time control, the process variable is often
called as point. A point can be a digital input/output,
a analogue input/output, a pulse input/ouput, a calculated variable, etc. For calculating convenience, we assume
all point are 4 bytes long, and identifier of a point is
2 bytes long(identifer of a point is used to identify a
data point in system range). Considering broadcasting
throughput provided by PROWAY, we have total updating
frequency

$$TUF=43K \ / \ 6 =7000 \ points/sec.$$

In actual execution, almost all the bandwidth be used to
broadcast(point-to-point data transfering is occasional),
so the TUF can be upto above 10000 points/sec.

We have discussed the priority deisgn of EPADM. According
the design, the individual update frequency of points is
actually the allocation of TUF. Suppose there are four
priority classes, the update frequency for each is U_1, U_2
, U_3 ana U_4, the number of update points for each priority
class in one second is X_1, X_2, X_3 and X_4, then we have

$$X_1 \times U_1 + X_2 \times U_2 + X_3 \times U_3 + X_4 \times U_4 = TUF$$

For PROWAY network, that is

$$X_1 \times U_1 + X_2 \times U_2 + X_3 \times U_3 + X_4 \times U_4 = 7000 \ points/sec.$$

The allocation of X_1, X_2, X_3 and X_4 in EPADM is done by
application user, so that user can use bandwidth ra-
tionally and well match the individual data update frequency to application requirements.

Here is analysis on a implementation based on IEC PROWAY
standard. Also we can do same analysis on implementation based on MAP standard MINIMAP, which will provides
much higher data update frequency.

From the analysis above, we have the conclusion. For
modern communication network can provide high broadcasting throughput, the data update frequency in EPADM is
acceptable by a common real-time control distributed
system.

Now let's look at second question. In EPADM, data are
stored in each station storage SS. The capacity of database is limited by the capacity of SS. Suppose internal
organization infomation for each point in source is S
bytes, in leakage is L bytes, each point has K redundancy
backups. Suppose the capacity of SS in each station is M
bytes, there are N stations in system, the update part of
a point is X bytes, then we have total database capacity

$$TV = M \times N \ / \ ((L+X)(K-1)+(S+X))$$

we have single database(local database) capacity

$$SV = M \ / \ ((L+X)(K-1)/K+(S+X)/K)$$

Suppose M = 512K bytes(this is not hard for modern
memery technology),N=30,K=4,X=4 bytes,L=6 bytes,S=
16 bytes,then

$$TV = 307.2K \quad points$$
$$SV = 40.9K \quad points$$

We can draw conclusion from above that the capacity of
database in EPADM is acceptable by common distributed
real time control system.

CONCLUSION

The goal of EPADM is to provide high processing effi-
ciency.The most advantage is its total processing
locality,its access response promptness.This is quite
suite to real-time control.Further more,EPADM also have
following advantages:

      .high reliablity and availablity.There is no
master station in EPADM,each local database damage
does not interfere other local database and global data-
base.Data is multiple backuped . All these bring high
reliablity and availablity.

      .Implementation simplicity.Comparing to tradi-
tional distributed database design,EPADM is much more
simple.For its data sharing and data broadcasting update
,and down of data distributity,the host access database
just like access a centralized database,no distributed
transaction processing,no all problems of distribution.
This make implementation simple,and well suited for
common industrial controllers.

      .Expandiablity.Adding of new local database to
global database is just like the new station adding to
the network (usually be done at same time),no interfer-
ence on other local database and global database.

Yet,EPADM adopts storage sharing and broadcasting data
update,the upbound of SS capacity and communication bandwidth limit the capacity of database and data update
frequency.Systems that have higher requirement for database capacity and update frequency than we gived above
may not apply to EPADM.

However,EPADM can be well used as a enhanced performance
design model of distributed database in real-time distributed control.It can be implemented directly upon
data link layer of network. We have developed a EPADM
experimental implementation on IEC PROWAY network.Execu-
tion proves its efficiency.The analysis data gived pri-
viously are come from this implementation.Further more,
the model is well matched to the new developed MAP
Enhanced Performance Architecture(EPA) of local industrial control network.

REFERENCE

.IEC PROWAY standard.
.Stefano Ceri,Giuseppe Pelagatti.Distributed Databases:
Principles and systems
.Proceedings of IFAC Workshop on distributed computer
control systems
.MAP/TOP communication network standards

INTEGRATING LOGIC AND
OBJECT PARADIGM IN HDDBMS

Zhang Xia, Zheng Huaiyuan and Yu Ge

Department of Computer Science and Engineering,
Northeast University of Technology, 1-1 Wen Hua Street, Shenyang, Liaoning, PRC

Abstract. The existing heterogeneous distributed database management systems (HDDBMSs) which are business-oriented will not be able to adequate for many new and developing information processing applications. The important objectives of these applications include the processing of complex objects and intelligence. Our approach is integrating logic and object paradigm to provide knowledge representation and inference mechanism in HDDBMS. This paper decribes the knowledge representation in object-oriented heterogeneous distributed database system with semantic association -- OHDDB-DG1, presents the method of processing constraints and rules in the global language OSAL which is Object-oriented and of Semantic Association with Logic, and explains briefly the peculiarities of processing inheritance and recursive inference.

Keywords. Distributed database; knowledge engineering; object oriented model; semantic association; semantic network.

INTRODUCTION

Because of the apparence of the relational database management systems, database technique has a great and rapid development and become an important branch of the computer area. For a long time, database management systems (DBMSs) have been regarded as a superior way for business data processing applications. However, as many new and developing information processing applications (such as AI, CAD/CAM, etc.) are researched deeply, it is now recognized that DBMS capabilities will be needed more. These applications will be characterized by the requirements of access to a variety of information types and access to substantial, possibly evolving, shared knowledge bases. Attempts to use existing DBMSs for these applications have generally failed because of lack of required functionality and performance. Hence, the subjects are attractive to research the DBMSs which would have the capabilities of processing complex objects, abstract data types, and intelligent inference.

This kind of subjects researches essentially the problems of combining knowlege based and database systems which have attracted researchers, both from Artificial Intelligence and Database. So far, the possibly developing strategies are basically three: (1). enhancements of existing systems (knowledge based or database systems), (2). coupling of independent knowledge based and database systems, and (3). design of a new class of systems, which are not constrained by the objectives and design characteristics of knowledge based and database systems. From the view point of the database area, it means design of a new data model which may integrate data and knowledge

representation.

The two formers present a short-term efficients. In the long run, the latter may lead to a final solution of integrating data and knowledge.

At the present, there are some researchers who have done a lot of researches using the latter strategy in the field, and there are some DBMSs such as PROBE, POSTGRES and Iris, etc. which have some developments in the aspect of presenting rules and inference, but there have been no mature experiences, all of them are in the stage of developing.

Object-oriented heterogeneous distributed database system with semantic association OHDDB-DG1 is one of the subjects in the development and research of CIM supported by the Chinese high technology project. Because CIM (computer integrated manufacturing) is the integration of BDP, CAD/CAPP/CAM and FMS, its data processing is very complicated, and the following features are required: the database is distributed in the physical location, and integrated in the logical concepts; it is heterogeneous, which have many kinds of DBMSs; it must support hierarchical recursive control and real-time control; it orients to complex objects (e.g. ordering, planning, financing and accounting, inventory, sale and market forecasting, etc. in BDP, graphics in CAD, robots, trays and tool bases, etc. in CAM and FMS); its data are of different structured levels and store in different media; and intelligence is also required so that the data or facts accessed in databases can be calculated, processed and even inferred to achieve intelligent control, decision aided and optimized manufacturing, etc.

Based on the above mentioned, OHDDB-DG1 is managed hierarchically, and seperated as basic database management system (BDMS) and distributed database management system (DDMS). The whole system has the database prism architecture, which is shown by Fig. 1, where INGRES, ORACLE and INFORMIX are relational sites, and DMU/FO is a distributed network database management system developed by our university. Its global data model is the Object-oriented Semantic Association with Logic model (OSALM). We have referred to Su (1983, 1986)'s SAM* in designing the structures of OSALM. OSALM can model the aspects of data distribution using generalization concepts, whereas most distributed database systems do not capture the partition relationships in their underlying data models. It can explicitly model horizontal partitioning, vertical partitioning and derived partitioning. OSALM can also maintain semantic integrity and consistency automatically. In addition to the above functionalities, the logic inference mechanism has been brought into OSALM in order to model data and knowledge uniformly. The global data language of OHDDB-DG1 is the Object-oriented and of Semantic Association with Logic data language (OSAL). It integrates logic and object paradigm to provide users with an advanced interface, to simplify the complexity of programming applications, and to enhance the opportunity to optimize queries to achieve the needed efficiency.

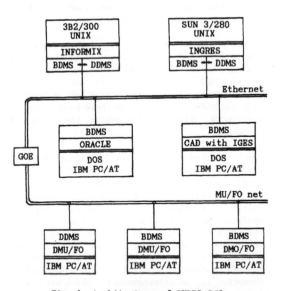

Fig. 1. Architecture of OHDDB-DG1

This paper is organized as follows. Section 2 gives the description of the knowledge representation in OHDDB-DG1: object paradigm covering semantic network. In Section 3, the knowledge representation language OSAL integrating logic and object paradigm is described and illustrated. Section 4 explains briefly the inference procedure. A summary and a conclusion are given in Section 5.

KNOWLEDGE REPRESENTATION IN OHDDB-DG1

Any method representing the mankind's knowledges must have two functions: one is to represent the facts, another is to represent the relationships among these facts, i.e. to get the information of another facts from some facts. This two functions may be implemented by using two different kinds of mechanisms. For example, the facts may be represented in the form of a series of predicate calculuses, and then the relationships among the facts may be represented in some form of indexes and sorting. But the semantic network presented initially by Quillion and Raphael in 1968 represents both of contents in the single mechanism. In the fact, the semantic network is the diagram representation of knowledges. Althrogh the semantic networks are very different in the concrete forms, their essential features are common. It consists of nodes and arcs connecting nodes, in which the nodes drawn as dots, rectangles or circles in diagrams represent the physical objects, concepts or situations, and arcs represent their relationships. In addition, the nodes and arcs may have labels.

In OSALM (Zhang, 1989), the basic element "object" is expressed explicitly by the structural properties, operational characteristics and constraint rules. Objects are grouped together into object classes based on some common semantic properties they share. An object can belong to more than one class and have a different representation in each of the classes. An object's representation in a class is also referred to as an instance of the class. The instance and class capture the semantics of "is a" (ISA). Object classes are divided into entity object classes (E-class) and domain object classes (D-class). The fundamental difference between the E-class and the D-class is that an E-class contains a set of instances, which are explicitly created by the user of the database, whereas, the D-class does not have any such instances, and is purely a domain specifying the data type or structure, a permissible range of values, etc. The D-classes are also seperated as simple D-classes and composite D-classes representing complex data types. In OSALM, the semantic and structural relationship between a class and some other class(es) is captured by an association of the class. There are two kinds of associations: semantic associations and data constructors which serve to define domains of complex data types such as Vector, Matrix, Set and Ordered-set, etc. The semantic association is the main carrier of semantic information in OSALM. There are five semantic associations, namely, Aggregation (A), Generalization (G), Interaction (I), Composition (C) and Cross-Product (X). The aggregation association defines a set of attributes for the defined class; the generalization association captures the semantics of "is a kind of" (AKO); the interaction association models the relationship among facts; the composition association aids in characterizing the set of objects of a constituent class as a whole; the cross-product association defines statistical category information. An entity object class must have at least one of the associations, it may be one of the A, G or I/C/X.

OSALM has the following features in the object-oriented aspect:

Encapsulation. OSALM models data and operation altogether. An object class consists of a specification part and an implementation part.

Object identity. An object is identified by a globally unique identifier (OID), which is generated by the system at the time of the object's creation. An object may exist independently of its values.

Classes and types. OSALM supports the concepts of classes and types.

Inheritance. OSALM allows objects with different structure to share the structure properties, operational characteristics and constraint rules associated with their common part by means of instances and object classes, subclasses and superclasses connected with generalization association.

Overriding. OSALM allows different operations to have the same name.

OSALM is itself a semantic network. It is also a diagram representation. Its semantic structural information may be represented in the semantic diagram (S-diagram), Figure 2 is an example of simple S-diagrams. OSALM has another powerful functionalities of processing complex objects and data types, and representing data distribution, etc, whereas, the general semantic networks haven't. In view of these facts, OSALM may cover semantic network-based knowledge representation in Artiffical Intelligence.

Concept nodes in semantic network, nodes which are used to express basic physical objects and concepts, is represented as entity object class in OSALM; Instance nodes in semantic network is referred to as instances of object classes; Link (or arc) ISA which connect a concept node and an instance node in semantic network represents membership, and instances in an object class is the member of the class in OSALM; Link (or arc) AKO connecting two concept nodes in semantic network represents the relationship between the subclass and the class, and the subclass-class relationship may be represented by means of generalization association; the other attribute links of nodes in semantic network are analogous to the attribute links of aggregation association in OSALM.

There is no formalized semantics in semantic network-based knowledge representation, that is, there is no uniform definition for the semantics presented by the given representation structure in contrast with predicate logic. The meanings assigned to the structures of semantic network are

determined completely to the properties of the procedure managing this network. There have already been some kinds of semantic network-based systems. Their inference procedures are different, but in common matching and inheriting are concerned in their inference procedures. In OSALM, there are the problems of inheritance and data matching for the structural properties, operational characteristics and constraint rules.

As OSALM is the global data model of OHDDB-DG1, the knowledge representation of OHDDB-DG1 has the form of semantic network from the view point of knowledge engineering. The global data model of OHDDB-DG1 may represent knowledge so that it is possible that the knowledge base sites are included in the heterogeneous distributed database systems.

THE ASPECT OF KNOWLEDGE REPRESENTATION IN OSAL

Since OSALM represents knowledge in the form of semantic network, a corresponding language OSAL for knowledge representation is required to implement its facilities. Furthermore, the logic paradigm is included in the OSAL.

In OHDDB-DG1, the extension of knowledge is a set of the facts stored really in the database, and is represented by the data model OSALM, the intension of knowledge is a collection of constraints (integrity control, protection, etc) and rules to deduce a collection of new data or facts from existing data.

Since the extention of knowledge is represented with OSALM, the users can use OSAL to define and manipulate it.

The constraints as the intension of knowledge which are operational generally can be represented with situation-action rules, which can improve and emphasize the performance of DBMS in many respects such as integrity and access control, exception reporting, alerting and notification, protection, etc. But most of existing DBMSs are limited in their ability to capture and process operational knowledge, which are only largely passive repositories of knowledge. Althrough it is always possible to store situation-action rules in databases, the DBMS does not spontaneously use this knowledge. Only when invoked by a user's query or transaction can the stored knowledge be used. Thus, an application that requires the use of operational knowledge must pose queries to the DBMS to evaluate the situation parts of its rules. Obviously, it would be more effective if the DBMS were active, i.e., if it continuously monitored the situation parts of the rules, and, whenever one was satisfied, it invoked the corresponding action immediately. A rudimentary active facility called as the trigger or alerter is built in OHDDB-DG1. The situation-action rules supported can be inputted in the definition section of OSAL, which have the formats as following:

```
CREATE  <Entity  class name>
         [INCLUDE <Local Entity class
                  name list>]
      <Entity class structure definition>
      [OPERATION <User-defined operation>]
      [RULE
         {TRIGGER-CONDITION ([INSERT,]
                            [DELETE,][MODIFY]);
            {IF <situation> THEN <action>;}
         END TRIGGER;
         }
      END RULE
      ]
      [AT SITE <site number>]
```

where [] describes the optional entry, {}
describes the entry which can repeats more
than once. A constraint rule consists of <situation>
and <action>, which are called precedent and
postcedent, respectively. The precedent describes
the condition and is represented by logic
expression; the postcedent   describes the
conclusion derived or action to be taked if the
condition are true, which can be a routine call
(e.g. displaying error messages) or assignation to
modify a value. In OHDDB-DG1, the constraints are
applied in insertion, deletion and modification to
a instance of the entity class, the action can be
executed if the condition is true.

In many systems such as POSTGRES (Rowe, 1987),
PROBE (Dayal, 1986) and Iris (Fishman, 1987),
rules to derive new data or facts from existing
data are represented as view definitions in their
query languages. Defining the views for the
intensional knowledge representation in DBMS have
the same role as defining the predicates by using
logic rules in AI language such as PROLOG. After
this kind of views are defined, a user can pose
his queries on the views, which are converted to
the query sequence which can be executed on the
stored data. The query modification technique has
the same role as deduction to resolve ask/question
problem in AI system. In OHDDB-DG1, we use this
kind of view definition approach to describe
rules.

In existing DBMSs, the crucial problem for
knowledge representation is that existing query
languages are too weak to capture all essential
intensional knowledge. Specifically, since most
existing query languages have the power of first
order predicate calculus, they are incapable of
expressing and processing resursion. No language
equivalent to first order predicate calculus can
express the transitive closure. As a result this
knowledge has to be embedded in application
programs, and is unavailable to the DBMS for
optimization so that it is impossible for some
rules to be inferred because of the long elapse of
time. Usually, recursion can arises in data
structures occurring in application programs.
While these recursive structures can be stored
extensionally in existing database (OSALM allows
direct and/or indirect recursive definition of
data structures), they can not be defined
intensionally as the results of queries.
Therefore, we improve the view definition and

query language capabilities of OSAL to include
recursion, and correspondingly extend the query
processing strategies of DBMS to process recursive
queries efficiently.

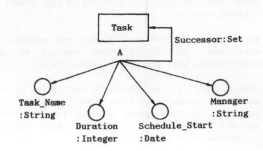

Fig. 2. Example of recursive definition

Suppose there exists a OSALM schema as shown in
Fig. 2 in the database. Now, we want to get the
information for all the tasks that might
potentially be affected if one task is delayed. To
do this, we define a view as following:

```
DEFINE VIEW Task_affected
   AS RETRIEVE Task_name, Successor
      CONTEXT Task;
DEFINE VIEW Task_affected
   AS RETRIEVE** Task_affected.Task_Name,
               Task.Successor
      CONTEXT Task_affected(Successor)*Task;
```

The view Task_affected is defined by two
definition commands including recursive
definition. In fact, this kind of recursion is
transitive closure (operator ** in the above
example is a transitive closure operator). After
the above views are defined, we may execute the
following query statement to check all the taskes
affected by task T1:

```
RETRIEVE Successor
CONTEXT Task_Affected [Task_Name = 'T1'];
```

To process this query, the existing query
processing algorithm must be extended with the
ability of generating recursive query commands.

                   REASONING IN OSALM

There are two kinds of OSALM reasoning processes
in OHDDB-DG1: one is inheritance, another is
recursion. We will briefly explain the features of
these two processes in OHDDB-DG1.

Inheritance

The inheritance in OSALM is a process of
transfering description of a object from its
object classes or its super classes to the
instance. This reasoning process is similar to
that of a person's thinking. Once he knows of the
identity of some object, he can associate many
descriptions about it in his mind.

The definition of object in OSALM is represented by structure properties, operation characteristics and constraint rules. Thus, the inheritance is related to structure, operation and constraint rules. When we make reasoning by the inheritance of OSALM, we use the breadth-first search strategy.

Recursion

One of the most crucial and difficult problems in the rule reasoning is recursion. The recursion calculus is essentially transitive closure calculus in DBMS. Therefore, the problem to find out a effective recursive query processing algorithm becomes the one to find out the algorithm of transitive closure calculus.

There are many algorithm of transitive closure such as Warshall's algorithm, Schnorr's algorithm, Dijkstra's algorithm, Topologically ordered traversal algorithm and Iterative algorithm. OHDDB-DGl use the Iterative algorithm, the essense is to execute an operation iteratively and compare the recult with the one of last time until the results are the same. The details about the algorithm are not introduced because of the paper length.

CONCLUSION REMARKS

The global data model OSALM of OHDDB-DGl can cover most semantic network knowledge reprentation of AI. The semantic network can be built on OSALM by means of object classes, generalization association, aggragation association. In this way, a basis has been set up to include knowledge base sites in heterogeneous distributed database system.

The global language OSAL of OHDDB-DGl integrates logic and object paradigm to represent and process the intension of knowledge: constraints and rules. The inclusion of constraints and rules make DBMSs from passive to active to improve the performance of the system greatly.

The effective processing of inheritance reasoning and recursion make OHDDB-DGl absorb many advantages of semantic network and logic expression and enhance its reasoning processing ability.

In a word, OSAL language has possessed some

reasoning ability, it can not only support knowledge processing in heterogeneous distributed database system, but also can be used to build up a knowledge base.

REFERENCES

Dayal, U., and J. M. Smith (1986). PROBE: A Knowledge-Oriented Database Management System. In M. L. Brodie, and J. Mylopoulos (Ed.), On Knowledge Base Management Systems, Springer-Verlag, New York. pp. 227-257.

Fishman, D. H., and others (1987). Iris: An Object-Oriented Database Management System. ACM TOOIS, 5, 48-69.

Mével, A., and T. Guéguen (1987). In M. Wolczko (Ed.), Smalltalk-80, MACMILLAN Education Ltd., London.

Rowe, L. A., and M. R. Stonebraker (1987). The POSTGRES Data Model. Proc. of the 13th VLDB Conf., Brighton, 83-96.

Smith, J. M., and D. C. P. Smith (1977). Database Abstractions: Aggregation and Generalization. ACM TODS, 2, 105-133.

Su, S. Y. W. (1983). SAM*: A Semantic Association Model for Corporate and Scientific-Statistical Databases. Information Sciences, 29, 151-199.

Su, S. Y. W. (1986). Modeling Integrated Manufacturing Data with SAM*. IEEE Computer, 19, 34-49.

Vassiliou, Y. (1986). Knowledge Based and Database Systems: Enhancements, Coupling or Integration?. In M. L. Brodie, and J. Mylopoulos (Ed.), On Knowledge Base Management Systems, Springer-Verlag, New York. pp. 87-91.

Wiederhold, G., R. L. Blum, and M. Walker (1986). An Integration of Knowledge and Data Representation. In M. L. Brodie, and J. Mylopoulos (Ed.), On Knowledge Base Management Systems, Springer-Verlag, New York. pp.431-444.

Zhang, X., and H. Zheng (1989). OHDDB-DGl: An Object-oriented Heterogeneous Distributed Database Management System with Semantic Association. Microcomputer, 9, 17-22.

INTERTASK-COMMUNICATION INSIDE A
REAL-TIME DATABASE

W. Stegbauer

*ABB Processautomation, PRIMO/S-Center, PA/ESS, Neustadter Straße 62,
6800 Mannheim 31, FRG*

Abstract. The implementation of process control systems
requires real-time features from the underlying database. Those
features are unknown in available commercial systems. There is
a demand for dedicated database systems that are welltailored to
the used hardware and operating system. Efficiency is much more
an issue here than portability. In a process control system
there is usually a large number of different, independently
scheduled programs (tasks) that access concurrently common data.
In addition, information has to be exchanged between the
cooperating tasks via well-defined interfaces. These interfaces
must be visible from outside. In the implementation described
in this paper, these interfaces were realized inside the
database which opens a huge field of possible applications.

Keywords. Data structures; Database management systems;
Industrial control; Process control; Program testing; Software
tools; Intertask-communication.

INTRODUCTION

In a wide range of process automation (e.g.
production control, building automation,
power plant control etc.) Asea Brown Boveri
(ABB) uses an own relational real-time
database called PRIMO/S, running under
VAX/VMS. The database was especially
designed for real-time applications and
meanwhile offers a wide range of
functionality including a Database Command
Language, a Data-Dictionary, Mask and
Report Generators.
In addition, PRIMO/S supports
intertask-communication inside the database
through data structures called Message
Arrays.
This paper describes the provided Message
Array functionality, the Message Monitor
and possible applications of Message Arrays
in a variety of fields including data
security and data replication in a
distributed environment.

PRIMO/S -
A RELATIONAL REAL-TIME DATABASE

PRIMO/S uses only the basic concepts of the
relational theory. On the other hand,
PRIMO/S meets the particular demands on a
real-time data management system which we
can be found among the following list of
arguments:

o A primary requirement on a real-time
  data management system is a very short
  access time (at least 10 times faster
  than commercial systems on the same
  hardware). All following points are
  subordinated to this.

o The size of the databases in a
  real-time data management system is
  relatively small. That is, a few MB
  up to 20-100 MB and for one relation
  not more than several tenthousands
  rows.

o The data structure in a process
  control system is stable at runtime.
  There is no need for online
  modifications of the fixed data
  structure. Of course, a real-time
  data management system also has to
  support the (off-line) change of data
  structures.

o In a Process control systems we have
  an extensive software package, which
  contains several tasks with common
  data and which depends to a large
  extent on intertask-communication. A
  real-time data management system has
  to cover this aspect (while commercial
  database management system do not
  offer anything in this area).

o The most frequently used database
  access is a random access to tuple
  data of one relation.

o In a real-time application, a large
  percentage of all accesses are read or
  modify accesses. Online deletions and
  insertions are rare.

o Graded data access niveau:

  - Fast logical accesses.

  - Faster processing through
    combined key/row number access.

– Very fast Direct Accesses for time critical applications (application programs can use for this purpose internal, physical knowledge, like knowledge about sorting of data or knowledge about physical pointers).

Those dedicated real-time features, especially the Direct Access, are like a scalpel: If you give it to the right person he (or she) can work wonders; if you give it to the wrong person it might end up in a mess.

These arguments constitute mostly restrictions with respect to a general purpose and most flexible solution. Special techniques can make full use of this knowledge for implementing highly efficient access algorithms.

IMPLEMENTATION CHARACTERISTICS

In the implementation of PRIMO/S a variety of techniques are used to fulfil the stated requirements:

o One-process architecture: PRIMO/S is shared common code. No (CPU-intensive) context-switch occurs when PRIMO/S Routines are called.

o ASSEMBLER was chosen as the implementation language for the PRIMO/S Kernel (while a portable PASCAL is the implementation language for the Utilities).

o The database itself is a shared region mapped into the address space of the calling program. This technique has the advantages of a memory-resident system (i.e. high data availability in core), while it does not suffer from it's shortcomings (i.e. memory is usually too small).

o A simple and very efficient locking mechanism was implemented. It uses only a few hardware instructions. In the non-conflict case only inline code is executed.

o The logic one-tuple-at-time through a call-interface was implemented.

o Transaction and recovery management are only available on explicit user request.

A typical random access (via a collision-free hash search) in a relation of 60.000 records requires approx. 400 MicroSeconds CPU-time (measured on a VAX-11/780).

INTERTASK–COMMUNICATION THROUGH MESSAGE ARRAYS

We decided to implement the intertask-communication inside the database via special relations called Message Arrays. (Relations in PRIMO/S are called arrays).

Message Arrays are like mailboxes. There are many possible Senders (=a task sending messages) and one dedicated Receiver which can be identified with the Message Array. The system synchronizes the Senders and Senders with the Receiver. The Receiver is automatically triggered when the first message arrives ('The postman rings once').

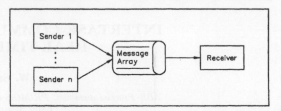

Fig. 1. The concept of a Message Array.

Message Arrays are 'normal' arrays with an additional maintenance of four parameters: RPI, RMC, SPI and FMC.

o RPI (Fig. 2) – the Receiver Pointer Index – identifies the row number of the next message to be received.

o RMC is the Read Message Count, the number of messages that have to be processed by the Receiver (in Fig. 2 the value of RMC is 4). RMC is by far the most inportant figure because it measures the amount of work that still has to be done, i.e. the load of the system.

o SPI – the Sender Pointer Index – identifies the 'oldest' message in the system, the message that will de destroyed during the next send.

o FMC – Free Message Count – identifies the number of rows that are currently free, i.e. available for a send call.

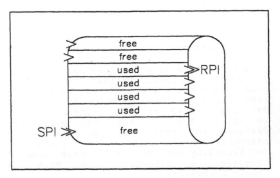

Fig. 2. Message Array realization

Message Arrays provide the following functionality:

o By default, the Receiver is stopped inside PRIMO/S, if the Message Array is empty. However, the Receiver can also (on request) regain control in this case and implement an own conflict strategy, e.g. waiting for keyboard input OR for message input.

o FIFO, LIFO and sorted sending and priority controlled sending are supported.

o Each access returns information about the actual Message Array status. Hence you can easily implement supervisor or tracing functions, bottle-neck detection etc.

o Message Arrays are 'normal' arrays in the sense that all query accesses are possible. In case of trouble all the database tools can be used for analysis of the history of the trouble.

Send/receive calls are approx. 5 times faster than message exchange via VMS mailboxes.

Message Arrays with Manual Deletion

The system provides an option concerning the deletion of messages. By default, a received message is automatically deleted and the occupied space is made available for new arriving messages.

However, PRIMO/S also supports manual deletion of messages. Manual deletion means that the received messages are put in a special area called Floating Area (Fig. 3): A message in this area is protected against Senders. The Receiver can either free messages in this area or the Receiver can reset the RPI to any message in this area and receive the same messages another time.

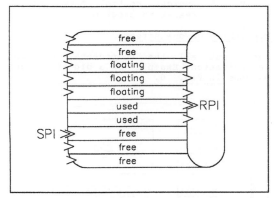

Fig. 3. Message Array with Floating Area

There is a variety of applications for this technique. If we look at a distributed environment where front-end computers close to the process peripherals pre-process indications and then mail them to one of the two host computers, the front-end process will usually ask for an answer from the host. If the indication messages are successfully stored on the host's disk, an answer is mailed to the front-end which in turn will free messages in the Floating Area.

However, if no answer is received in a certain time and after several communication attemps, the front-end process resets the RPI to the begin of the Floating Area and starts sending the indication messages to the second host (which is a hot stand-by).

MONITORING MESSAGE ARRAY ACTIVITIES

If we are able to look at the amount and contents of exchanged messages we see what happens in the whole system. This is the simple background philosophy of the PRIMO/S Message Monitor which offers the following functionality:

o Show overview on the status of each Message Arrays. For each Message Array, the four associated parameters RPI, RMC, SPI and FMC are displayed. In addition, you can see if the Receiver is actually running (i.e. "/ru" in the column "Receiver" in Table 1) or waiting ("/wa")

o Flag all recognized changes on screen by an appropriate video attribute. You watch the message running through the system.

o Send/Receive messages, with optional repetition count and/or with optional delay time. This means you can actually test completely the interface to the Receiver without having any Sender available: You store all the (specified) types of possible message in a test file and let the Monitor send them – and check the Receiver's reaction.

The Monitor is an important help during program development, single module test and simulation of hardware environment. It is the tool for integration, tuning, bottle-neck detection and trouble shooting in the running system.

Table 1 gives an idea of the Message Monitor screen output. We see a list of Message Arrays and for each one the characteristic actual values. The most important value is RMC, the Read Message Count (displayed in two ways, as an absolut value and also as percentage of available space). RMC shows the amount of work the Receiver still has to do.

TABLE 1   Message Monitor Display

Array	RMC	RPI	SPI	FMC	Receiver
INDI	0/0%	7	7	100/100%	hib/wa
DISP	6/2%	1	7	294/ 98%	hib/wa
PROT	3/0%	4	7	997/100%	hib/ru
ALRM	0/0%	3	3	25/100%	hib/wa

Table 1 represents an example from real life, every figure is meaningfull and carries a lot of information. We extract the follwing facts:

o The INDI and the ALRM Receiver have done their work and are waiting for new work (both RMCs are zero and Receivers are "/wa"iting).

o The INDI Receiver has processed 6 messages, ALRM has processed 3 messages.

o The DISP Receiver has 6 messages to process – but he does nothing. Since RPI has still value 1 it is very likely that this Receiver was never started during the lifetime of the system.

o The PROT Receiver has some work to do and is actually "/ru"nning – a consistent combination (as it is in real life). It is processing the third message and still has three messages to process. We expect that this RMC will be zero in near future, i.e. in the next output cycle of the Message Monitor.

USING MESSAGE ARRAYS IN A PROCESS CONTROL SYSTEM

In our model of reality, each single process in a process control system can be seen as a black box that interfaces with the outside world via Message Arrays or hardware (like a key board, a screen, a printer or the process peripherals). We now elaborate this point of view in an extensive example.

A Demonstration Example

Let us consider in more detail a possible database application in a process control system.

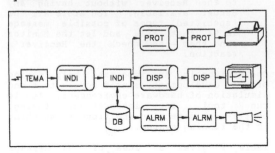

Fig. 4. A typical process application using Message Arrays.

On the left hand side of Fig. 4, we have the telegram manager TEMA which interfaces with the process peripherals. TEMA receives telegrams which contain (for the sake of our example) an address and the status of an indication, i.e. the status of a switch or a pump or a valve. TEMA passes the message to the indication manager INDI for processing. INDI reads the complete actual database information for this indication and updates the process image in the data base.

INDI then starts the analysis of a bitmask for that indication as it was read from the database. In presence of an appropriate bitsetting INDI passes the indication event to the protocol manager PROT, to the display manager DISP or INDI triggers an acoustic ALARM. Table 1 represents a possible status of each Message Array in this environment.

If nothing happens in the controlled peripheral process, all tasks wait for an event. If an event occures it is passed through the Message Arrays. If an indication burst occures the Message Arrays fill up to a certain level until the system again calms down.

Control Flow and Call-Interface

We can see from our example: Unlike batch processes (that are started, do their work and finally exit), real-time processes are active and hot at all times as long as the system is alive. The typical program control flow in real-time processes is outlined in Fig. 5.

The PRIMO/S call-interface is especially designed with respect to this control flow: A typical data base call is prepared in an initialization phase ('Construct'), the task waits for an event, it processes the event and then again it waits. The same constructed calls are used again and again in the processing loop.

The prepared calls are validated and optimized in the phase 'Construct' and efficiently used as binary data structures at processing time. The checking of names, the computation of addresses, the query optimization etc. are all done at construct time and stored in a special internal call-specific data structure that is referenced at processing time.

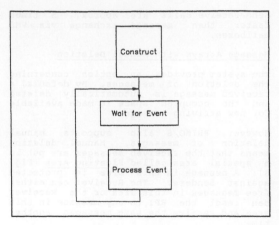

Fig. 5. Typical control flow in a real-time program.

Everything on one screen

Let's put all the components of our Demonstration Example on a big 19" screen as done in Fig. 6.

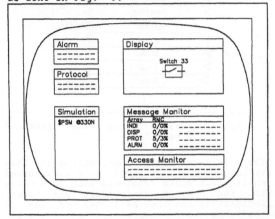

Fig. 6. The whole system on one screen.

In the upper part of screen we recognize the output components of the running system. In the lower screen part we simulate the process peripherals and we monitor the data base activities.

We start playing with the system by a simple simulation of an incoming indication (by an appropriate command in the simulation window we put the displayed switch 33 to status ON). We see through the Message Monitor which Message Arrays are affected and if the message is passed properly. We see the output effects on DISPLAY, on PROTOCOL and on ALRM, respectively. We can consult the Access Monitor (another available PRIMO/S Utility) and answer questions like "How many data base accesses are involved when processing one indication?"

We can also estimate the efficiency of the system. How many indications can the system process in one second? We use again our Message Monitor and send through the simulation window a certain indication, say 1000 times with a delay time of 10 MilliSeconds. We can see what happens on the Message Monitor screen display: Are all tasks efficient enough to handle the event burst? Or are there bottle-necks?

We can actually see: Bottle-necks are
those Message Arrays with increasing number
of unprocessed messages (=RMC). This means
of course that the corresponding Receiver
is the bottle-neck.

FURTHER MESSAGE ARRAY APPLICATIONS

The Message Array technique opens a huge
field of possible applications. Some
further ideas were born on the application
front during the realization of big
projects and are outlined in this section.

System Error Logging

Each process control system must have a
philosophy about handling recognized
errors. We have the following.

All system processes send their detected
errors into a Message Array, say SYSLOG.
The Receiver of SYSLOG timestamps the
message and does further processing like:
writes it to a file, controls available
disk space, classify and counts errors,
informs host system, informs operator etc.

Like any Message Array, SYSLOG can be
online inspected via the Message Monitor
and also via interactive database
utilities. Questions like 'Was there an
error during any database access in the
last hour?' can be answered online.

Separation of Keyboard Input
and Screen Output

In an interactive display application in an
process automation system it usually occurs
that messages have to be displayed on
screen, although the operator does not
enter any keystroke. In the simplest
example the actual time on screen has to be
refreshed every n seconds.

Conventional solutions to this problem
include asynchronous procedures or brutal
break through writes on the screen by
foreign processes.

PRIMO/S provides a more elegant solution to
this problem: The keyboard input is
received (via a simple program) and send to
the Display Manager via a Message Array –
for possible echo and further processing.

All other processes can inform
synchronously, again via this Message
Array, the Display Manager about system
events (e.g. errors, problems, tracing
states, troubles on disk space or with
printers etc.)

Keeping a slice of the past

In an nuclear power plant application it
was a requirement, that a certain amount of
indications, say the last 10 thousand, must
always be available. In case of troubles
these indication list was analysed for
error conditions in the past.

We used a Message Array with the OVER RUN
option to implement this feature. There
was no development effort besides simply
sending each indication in this Message
Array. The OVER_RUN option guaranties that
the oldest message is deleted if message
number 10.001 comes in. In this way, a
piece of past is provided at all times.

Controlling the Receiver

If we look at the Receiver of a Message
Array we can easily drive him to change
internal states like:

o 'Put on/off test level'

o 'Rewrite your output file now',

o 'Show internal states or variables',

o 'Change your internal processing
   strategie'.

o 'Clean-up and Exit',

All these control messages, which can be
identified by a certain message type, are
included synchronously in the stream of
'normal' application messages for this
Receiver.

MESSAGE ARRAY INTERFACE
FOR COMMAND LANGUAGE

PRIMO/S provides a database Command
Language called PSU. PSU accepts
interactive input as well as input from
batchfiles. Wouldn't it be a nice idea to
have PSU interfacing with a Message Array
as input? It took a development effort of
one day to implement such an interface as
seen in Fig. 7.

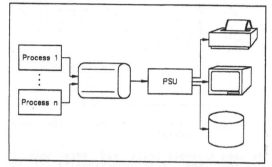

Fig. 7. Command Language PSU with
        Message Array interface

The advantages of this realization are as
follows:

o PSU is one server that can have an
   arbitrary number of clients in the
   system. This helps to reduce the
   number of concurrently active
   processes.

o PSU is active at all times, PSU has
   the database(s) opened and can be
   triggered by a fast mechanism.

o The full functionality of PSU is
   available for all programms.

IMPLEMENTING SQL
ON TOP OF PSU

SQL is on everybodys favorite list of
database command languages. Our (long time
existing) PSU provides nearly the same
functionality. So why don't we translate
SQL-statements into PSU-statement and let
PSU do the work? And how do we pass the
translations to PSU? Again Message Array
was the key word to the implementation
which can be seen in Fig. 8.
The interface to SQL-functionality was
called PSQL. PSQL controls the keyboard
and echoes input in the lower part of the
screen. It translates a SQL-statement like

```
        SELECT *
        FROM books
        WHERE price BETWEEN 47.10 AND 49.10
```

into PSU-syntax (which is assumed to be self-documentary here)

```
        ARRAY books
        ATTRIBUTE *
        QUERY price >= 47.10 AND -
              price <= 49.10
        TYPE ALL
```

PSQL releases screen control and passes the translation as multiple messages to PSU and waits on a second Message Array for an answer of PSU.

PSU processes the messages and uses the upper part of the screen for output. If all work is done, PSU releases screen control and re-triggers PSQL which in turn then will accept another SQL-statement from the user.

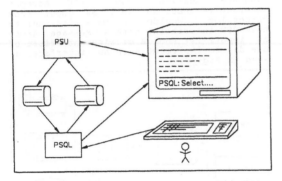

Fig. 8. Using PSU for SQL-imple-mentation

We used a simple trick and were able to provide the full functionalty of PSU through PSQL and it didn't cost us any development effort: When PSQL recognizes an unknown token in the keyboard input stream, it simply passes the token to PSU for processing. So all the PSU-specific command key words will automatically reach the PSU and properly processed.

USING THE MESSAGE ARRAY CONCEPT FOR DATABASE JOURNALING

It is an option in PRIMO/S that it uses a special Message Array JOURNA for logging all database updates. The provided functionalty simply states, that there is a availability of all updates in JOURNA - as can be seen in Fig. 9. The updates caused by processes P1 .. Pn are done in the online database and in addition journaled in JOURNA. Nothing is said so far about a possible consumer of JOURNA messages.

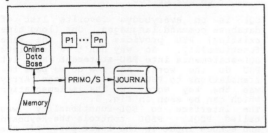

Fig. 9. The concept of Journaling.

Logging

The Logging Manager is one alternative as Receiver of JOURNA. It writes the received updates in a logging file. If the system fails, all logged updates are redone on a saved database copy.

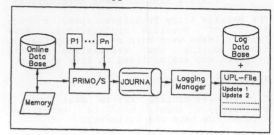

Fig. 10. The Logging Manager

Shadowing

The Shadow Manager is another possible consumer of the updates journaled in the Message Array JOURNA. The Shadow Manager maintains two copies of the online data base and reproduces the updates in both copies, one after the other. In case of a failure, at least one of the two copies is consistent.

Distributed Replication

The Journaling of updates in a Message Array can also be used to replicate all updates in database copies in any number of nodes (e.g. workstations, substations, front-ends etc.).

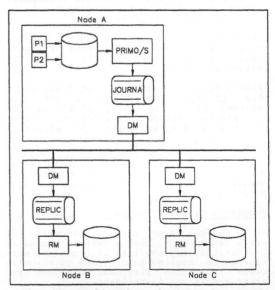

Fig. 11. Replication of data in remote nodes.

The processes P1 .. P2 in Fig. 11 are sources for updates - PRIMO/S journals them in JOURNA. A (simple) program called Distribution Manager (DM) receives the messages and has access to a configuration table that tells DM on which nodes the updates have to be replicated. It sends them via the inter-computer communication (like DECnet) to the remote nodes where a coporating DM receives the updates and

writes them into a Message Array called
REPLIC. The Replication Manager RM finally
replicates the update in the remote
database.

Replication in this implementation follows
the simple philosophy: Remote nodes are
updated as-soon-as-possible (it is the
author's belief that simple ideas - and
only simple ideas - work).

CONCLUSION

The data management system as outlined in
this paper has prooved in many projects to
fulfil real-time requirements. The
provided functionality, especially the
functionality of Message Arrays can solve
hard and well known problems in real-time
programming. In addition, the system
offers a software development environment
and tools for testing, monitoring, trouble
shooting and bottle-neck detection.

The functionality of Distributed
Replication will be extended in near future
and cover a true Remote Update in a DECnet
environment.

CRDB — A REAL TIME DATA BASE FOR
DISTRIBUTED SYSTEM WITH
HIGH RELIABILITY

L. Tapolcai and Z. Sopronfalvi

*Computer and Automation Institute, Hungarian Academy of Sciences, 1111-Budapest,
Kende u. 13/17, Hungary*

Abstract. This paper presents a distributed data base management system for serving high reliable supervisory control systems based on local area network. It is supposed that the distributed control system is built of the hosts which can replace each other and the database management system is provided for services which aid the organization switch on function from the breakdown host to the working host.

Keywords. Real-time database, distributed database, system with high reliability.

INTRODUCTION

Some supervisory control systems have to be at disposal with high reliability, for example: supervisory systems of power stations.

Formerly a system architecture of high reliability was achieved by defining on-line and stand-by machines and the whole supervisory system was running on the on-line machine, while the database was repeatedly saved to the stand-by machine through a very speedy, typically parallel line.

There was a problem to connect the device with a serial port. A special device was used to switch serial devices from one host to another, but the management of this device was very complicated.

In such a case, the stand-by was continuously ready (hot) to take over the on-line tasks, while its power and capacity was unutilized.

The possibilities offered by the LANs suggest much more flexible solutions. The functional tasks are uniformly distributed among host computers considered to the LAN. In this concept, the capacity of the host computers is available continuously and completely. In case of faulty operation in any of the host computers, a reconfiguration is activated immediately to redistribute the functional tasks among the available machines. Loosing more and more host capacity, less important functions can be altered to lower priority levels or even should be omitted.

The realization of his philosophy requires conditions to dynamically allocable functions. Database handling systems known by us, did not support this dynamic allocation. This is what motivated us to develop the CRDB (Core Resident Data Base).

Point of view of data handling

The technological environment is closed and it is changed very rarely. The changes are simple, they never have a revolutionary character. So supervisory systems can be regarded as target systems. All processes are predefined and they have a co-operative attitude. Cooperative attitude means that the processes are supported to operate the fair-play, no special efforts required from the database system to check their activity and their access to the database.

All the data structures and operations with them, as far, as all the aspects of the required search operations are well defined, no ad hoc situation is expected.

The database handling is slightly simplified as the key fields never change on-line.

The speed of record access is one of the crucial aspects in supervisory control systems which is of utmost importance in case when technological failures occur. The fast data access may be primarily achieved by using memory (core) resident database. The average size of a supervisory control system's database is 4-10 Mbytes which can easily be provided by the commercially available hardware.

These point of views were regarded by the design CRDB, too.

Dynamic reconfiguration of the system

Two special facilities of the CRDB will be described:

- servicing of the dynamic reconfiguration,
- supporting of the process-process communication.

By the design of the system top-down, the tasks are divided into functional groups, it means subsystem. Each subsystem many consist of one or some databases and processes.

Namely, the complete database, called the project database, is divided into separate parts. There may be 32 databases, each consisting of at most 64 files.

A database has three different states:

- ON-LINE
- SHADOW
- OFF-Line.

The whole run-time system may contain several copies of the subsystem and inside of the subsystem the particular database, but – at any time –, only one copy of them may be in the state ON-LINE.

Files in the database, in state ON-LINE, can be

71

read as well as written.

In the run-time system there may exist several
copies of each database in state SHADOW. These
database copies are regularly updated on the basis of the ON-LINE copy. The update frequency
should be determined for each individual database and file. File in database, in state SHADOW
can be read only.

Database copies in the OFF-LINE state are
"switched off".

Figure 1 illustrates the different database states
and their significance in rearranging the functional groups.

secondary database on Host B.

The processing programs can access the database in
two different ways:

- ON-LINE access: it can read, as well as
  write the copy of the database being currently in the state ON-LINE.

- CHEAPER access: it may be a passive operation
  from the database copy accessible at the lowest cost. If there is a copy of the database
  in state ON-LINE on the host on which the
  read operation is issued, then that copy is
  read. If there is a copy of the database on
  the host in state SHADOW on which the read
  operation is issued, then that copy is read.

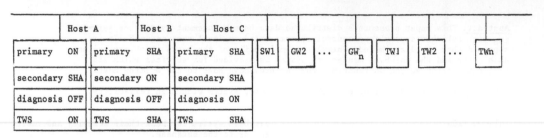

Fig. 1. Normal state of the system

The system consists of four subsystems and the
project database consists of four databases:

- The primary database contains data directly
  connected to data acquisition and required
  by the primary processing of the technical
  data, such as constants for conversion into
  engineering units, the values themselves
  obtained by measurement or signalling, etc.

- The secondary database collects the derived
  technological data, such as algorithms and
  constants used in computations, the result
  of computations, etc.

- The diagnosis database collects data, such as
  the knowledge base of systems diagnosis, intermediate results, active pieces of "ad-
  vices", etc., necessary for the diagnosis
  of technology.

- The technological workstation service database
  (TWS) contains data connected to service the
  functions of a technological workstation which
  is connected to a LAN, too.

The technological workstations (TW) realize communication between the operator and the system.

The gateways (GW) connect telemechanic lines to
the system.

Normally:

Host A   performs primary processing with the pri-
         mary database copy being in the ON-LINE
         state. The TWS database is being in ON-
         LINE state, too.

Host B   performs secondary processing with the
         secondary database copy being in ON-LINE
         state, and

Host C   performs the diagnosis using the diagnosis
         database copy being in ON-LINE state, too.

The copies of the primary and TWS databases on
Hosts B and C in state SHADOW are updated on the
basis of the primary database on Host A. Similar-
ly, the secondary database on Hosts A and C in
state SHADOW are updated on the basis of the

And if there is only an OFF-LINE copy of the
database on the host on which the read operation is issued, then the ON-LINE or SHADOW
database copy accessible at the lowest cost
is read.

In this case the processes of a primary subsystem
access primary database which is in state ON-LINE.

The position of the processes of the secondary
subsystem on Host B is more complicated.

These processes access primary database used for
calculating derived technological data. The access to primary database may be CHEAPER ACCESS
and the access to secondary database should be
ON-LINE. If they want to send data onto the technological workstation for the operator, they have
to execute ON-LINE access on the TWS database.
This operation will be executed remotely.

The access mode (namely ON-LINE or CHEAPER) will
be defined at the design time.

That the access will be executed locally or remotely, it depends only on the run-time status
of the subsystems. The system has a fifth subsystem, it is called by us: "supervisor". It runs on
each host. The supervisor checks the state of the
host, hardware and software, and if the system has
some troubles, the supervisor subsystem will execute the reconfiguration of the system. It means
some subsystems will turn off, some subsystems
will turn on.

Let us see the case, for example, when Host A
breaks down, it may be caused by sofware or hardware.

It is more simple if the whole Host A breaks down
because the subsystems running on Host A have not
to be turned off. In case, if only some parts of
(hardware or software) break down, the supervisor
turns off the subsystems on Host A.

After that, the supervisor turns on the subsys-
tems-primer, TWS - on Host B. Databases of these

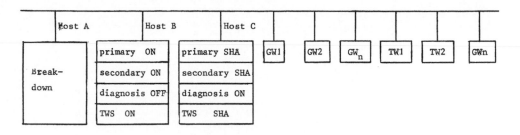

Fig. 2. State of the system after reconfiguration

two subsystems were in state SHADOW, so these were updated. The processes of subsystems will be turned on, too.

This time, the supervisor reconfigurates the communication with the gateways and the technological workstation, too.

Of course, to realize this method of reconfiguration, the processes have to be built in a special way.

This new system state can be seen on Fig. 2.

Support of the process-process communication

One of the most delicate problems of distributed control systems is the process-process communication. The solution of that problem is especially important and difficult if the system functions are distributed. The CRDB supports the solution by providing special FIFO queue files.

The FIFO files propagate the records to the consumer process either in the order they are received of in some other, previously specified order. The servicing of the servicing of the technological workstation is a good example to illustrate the usage of files of this type. Refreshing information can come at any spot in the distributed system and is sent to the FIFO type file TWFRESH in state ON-LINE of the TWS database. If Host A contains the currently ON-LINE copy of the TWS database in the above example, then fresh records on Hosts B or C are automatically gathered on Host A, and the TWS subsystem has link with the technological workstation which will send refreshing records to them.

Process-process communication is supported by a flag-technique, which can be assigned to the FIFO file filling procedure by the user. The fact that the number of unread records in a FIFO file exceeds a previously set limit is signalled by a flag. The producing processes should wait if the flag is set on. As soon as the number of records in the file becomes less than the limit, the flag is set off. This technique provides a simple and efficient way for synchronizing the producing and consuming processes and for organizing data flow.

One more facility of the FIFO queue files is the sort capability. You may define primary key for the FIFO queue file and by the write operation, the records will be sorted. It is very useful, for example, for gathering information with time attribute which comes from different parts of the distributed system and we want to see them in raising order.

The CRDB has special tools for monitoring the FIFO type files. It helps the programmers to ob-

serve and control data flow between the system components. It is very useful to tune the system.

CONCLUSION

A distributed supervisory control system for a high performance power station was implemented using CRDB.

The database management system proved to be very effective. During program development, the testing and debugging made it easy and efficient to test the individual subsystems as well as the communication between the subsystem.

During operation, the fast database switching facility made it possible to redistribute the vital functions to an operating host, thus providing a very high level of the availability of the system.

A DISTRIBUTED DATA BASE FOR
REAL-TIME CONTROL/MONITORING OF
POWER PLANTS

J. B. Lewoc*, E. Ślusarska** and A. Tomczyk**

*ZE Elwro, Wroclaw, Poland
**Institute for Power System Automation, Wroclaw, Poland

Abstract. Real-time monitoring and control of a power plant delivering 1000
MW or more power is a complex task. It involves monitoring of 10000 or so
analog signals and even more two-state signals. In addition, some data is to
be interchanged between the real-time systems and the off-line ones which are
used by the power station staff involved in run preparation functions. The
paper describes the basic hardware and software tools designed for development
of the distributed data base for a new power plant in Poland, which is a pilot
implementation of such data bases.

Keywords. Control system, management system, power generation, computer
network, performance bounds.

INTRODUCTION

The Institute for Power System Automation (IASE),
Wrocław is involved in computerization of power
plants in Poland. The most complex project is in
progress for the power plant Opole under
development. The power plant is to consist of 6
power generating units (PGU), 360 MW each, plus a
virtual PGU which is to be, in fact, a
configuration of environment monitoring/protec-
tion equipment vital for the power plant area
including the city of Opole.

IASE (1986) develops the direct control systems
for PGU-s including sequential controlles for PGU
switching on and off, turbine controlles which
are called here "sequential" controllers (SC).
More complex tasks are performed by PGU monitors
which were, initially, intended for PGU control
engineers'aids.

After the very initial phase of the project, it
was recognized that it is worthwile and, at the
same time, feasible to develop a distributed data
base for the power plant, using the computer
systems developed for individual purposes.

The data base Badel (from Polish Baza danych dla
elektrowni - data base for power plants) should,
inevitably, cover also the power plant dispatch
monitor and the off-line service computer network
though these components of Badel are being
developed by other contractors.

The paper presents the logical design of Badel
and the set of services to be provided for the
phase 1 to be in operation when the first PGU is
synchronized with the power network of our
country.

HARDWARE ENVIRONMENT OF BADEL

The hardware environment is shown in Fig. 1.
The 1-th PGU ($1\in1,..,n$) is equipped with 1_s SC -
type control systems which are implemented on the
Master computers (IASE (1986))based on Intel
8080A. The 1-th PGU's monitor is built on k_1 data
concentrators (DAQ) also implemented on Master
and two MERA 680 minicomputers (Meraster (1988))
compatible with the Digital (1982) family LSI-11:
the control processor (CP_1) and the data
presentation processor (PP_1).

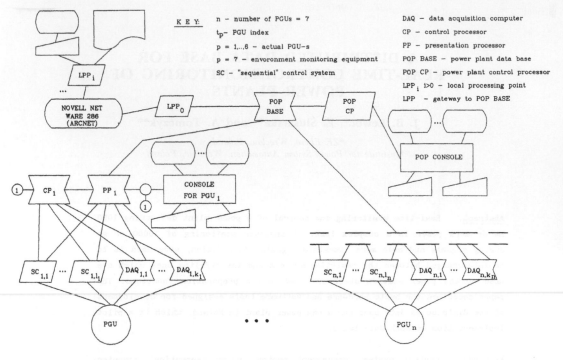

Fig.1. Power plant automation and management network (Badel)

The lower level micros and the higher level minis are interconnected via RS 232C (V.24) links operating, primarily, at the 4,800 b/s transfer rate.

All PGU monitors are interconnected with the MERA 680 computer in the multiaccess configuration. Here, RS 232C links are used too. However, the communication subsystem is provided with a front-end processor connected with the POP BASE host via a direct memory access (DMA) channel.

The power plant dispatch central processor (POP CP ; also MERA 680) and the local area network (Arcnet run under Novell (1986) NetWare 286) with one working station serving the functions of a gateway to POP BASE complete the Badel hardware.

LOGIC STRUCTURE OF BADEL

The logic structure by which we mean here the basic data files, data interchange protocols and retrieval functions will be described in the bottom to the top way, just as our solutions were designed in practice. The opposite direction, from the to the bottom has never been resulted in a successful implementation of a complex computer

system in our country. Some discussion of this problem is given elsewhere (Błach (1989)).

DATA STRUCTURES ON THE LOWER LEVEL

The basic data structures maintained in DAQ-s are shown in Fig. 2.

The structures are held in RAM and only one value of any type of a variable (current value, filtered value, a bit in the PGU topology) is stored. This is inconsistent with some popular solutions where the low level computers collect some history of the variables and current values of the latter are not, in general, reproduced on the higher level. The primary excuse for such solutions is the lower throughput demand for the communication subsystems. We consider the excuse a poor one since:

 - in normal operation there are still unused passbands in communication links while in abnormal conditions the 'economical' solution implies serious, unnecessary traffic and the overall performance is considerably worse,

 - even for such complex plant as a PGU, the monitor computer (CP) load implied by transmission and archivation of data does not exceed 15% when all information is sent to the

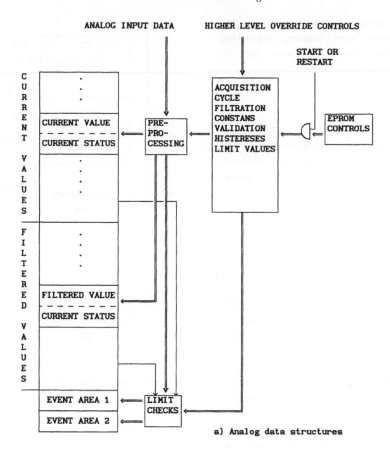

ANALOG INPUT DATA   HIGHER LEVEL OVERRIDE CONTROLS

a) Analog data structures

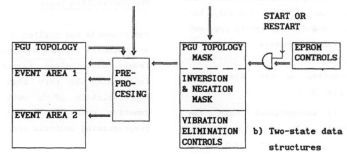

TWO-STATE INPUT DATA   HIGHER LEVEL OVERRIDE CONTROLS

b) Two-state data
    structures

Fig. 2. Basic data structures on the lower level

higher level in cycles defined by the process control needs,

- storing of current data on the higher level results in much higher fault tolerance (a DAQ failure may be "hidden" due to the data kept on the higher level) of the whole system and CP may be a source of "safety" data for every user; in the other case, rather considerable transfer delays may result in unsynchronisation in the some data but used by varions users.

The above premises were checked for the case of the 360 MW monitor network.

THE TRANSMISSION PROTOCAL WITHIN THE PGU MONITOR NETWORKS

General

Of the 7 layer level, model defined by ISO (1982) for the Open Systems Interconnection , the physical layer, link layer and application layer are implemented in the monitor network protocal. The design criteria for the latter were the efficiency (required due to severe throughput limitations) and legibility needed to ensure easy

implementation, operating and expanding the
networks. Therefore, more complex, automation
oriented protocols like TOP were to be
eliminated.

The Physical Layer

The physical layer protocol is defined by V.24
(RS 232C). Transmission is transparent and the
traverse parity is coded/tested. The protocol
does not distinguish transmission entities except
of bytes (octets).

The Link Layer

This layer is responsible for:

- distinguishing of transmission entities
called frames,

- transmission data validation realized with
use of the longitudinal checksum which, together
with the parity checks of the physical
level, ensures that the error omission probability
is not greater than ca. $2 \cdot 10^{-5}$,

- transport functions consisting in testing
the sender identity and, if OK, delivering the
frame to the application addressed,

- flow control ensuring that the high nor low
level computer input capacity may never be

exceeded (we eliminate completely the chance to
miss any octet due to software delays), the
requirements for data acquisition cycles are met
and transmission delays for random (unfrequent)
frames are possibly low.

For the above purposes, we use the frame
structure shown in Fig. 3.

The flow control is accomplished with the
command/reply field. The protocol is
unsymmetrical in accordance with the requirements
of automation applications and most commands are
issued by the high level. All commands need some
answer. Till now, only one command is implemented
on the low level: an event report substitutes any
frame requested by a command; the report itself
is also a command to ensure acknowledgment for
the frame and thus reliable sensing of events.

For the power plant under consideration, all SCs
and DAQ-s of one monitor are divided into 4
groups within which transmission is serial. For
DAQ-s serving up to 192 analog variables each,
and a transfer rate of 4800 b/s, proper
arrangement of the input system ensures meeting
all demands on communication subsystem
performance.

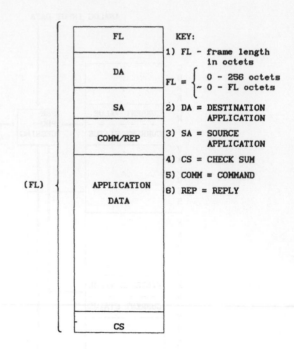

Fig. 3. FRAME STRUCTURE

The Application Layer

The commands and replies of the application layer
are mutually aggreed between the lower and higher
levels. Nevertheless, to ensure legibility, we
implemented the rule that each command is a
representation of a well defined application
function such us: Send topology, Receive
preprocessing controls etc.

BASIC DATA STRUCTURES
ON THE HIGHER LEVEL

These basic data structures are presented in
Fig. 4.

The kernel structure in this data base is the
process picture (PP) which is used to derive all
major files on the higher level. Let us mention
the history file which maintains the movie copy
of PP for some 30 minutes. If a major incident in
the PGU happens, the movie copie may be dumped to
one of the pre-incident and (30 minuts after) to
the correspondent post-incident one for analysis
after the emergency is cleared).

PROCESS PICTURE

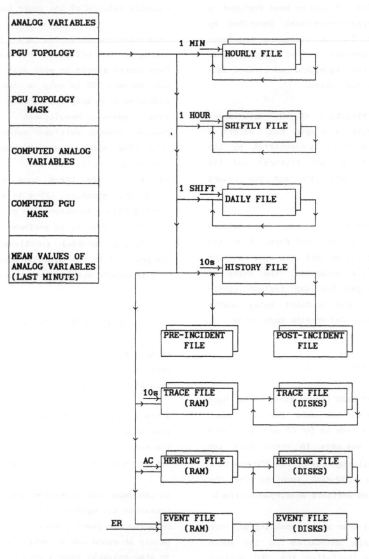

KEY: 1) AC = Acquisition cycle
     2) ER = Event report

Fig 4. Basic data structures on the higher level

The trace and herring files are (in a sense) subsets of the history file but arranged so that it is easy to follow courses of some variables (the traces are updated once per 10 s while the Herrings - in standard acquisition cycles).

Disk files are of cycle character (except of the incident ones). The data access methods are not too complex, but the application puts more serions demands on data retrieval services.

DATA RETRIEVAL SERVICES

The basic data retrieval services include:
  - individual processing of data (such as non-linear conversion to physical values, computation of trends, integrating etc),
  - preparation of VDU pictures (tables, schemes, plots etc),
  - preparation of reports.

To speed up preparation of the above, individually defined services we have designed a computer aided design system Cadel described by Rozent (1988),which is, de facto, a knowledge base enabling for process control engineers to edit (not to program) algorithms and control data required for individual cases.

The well defined structure of the data model enable us to apply a simple but effective solution. Each data entity is identified by its number (as for the process picture) and its location (ie. the data file and the record number).

Simple auxiliary procedures access the entities and convert them to the standard form. Thus the retrieval services may be used with ease on data kept in any file (this means that the state of the PGU may be analysed for any instant within the tracing scope of the monitor) using mostly the software prepared for on-line monitoring.

THE POP-BASE

The computer of POP-BASE (power plant data base enabling access to on-line data for all power plant users) is the same as for CP and PP: MERA 680. This decision was made in order that the basic data retrieval services of the PGU monitors can be the base for POP-BASE services for the phase 1 with limited software development effort.

POP-BASE is a multi-user system enabling any user to get access to data maintained in the whole Badel. There is one limitation: the PGU control engineers possess absolute priority to their data over any other user (eg. administration staff).

We assumed that there is no need to reproduce all PGU monitors'data in POP-BASE (the task is not feasible without high speed communication media like those of LANs). Thus, we transfer from the monitors to POP-BASE all event file records while the records of all other files are transmitted at the most once per minute. Data of better time resolution is delivered on demand only.

Except of the PGU monitors, the users of POP-BASE may be (Lewoc (1989)):
    - any power plant user having access to the local area network,
    - the power plant dispatcher,

    - any user having access to the future computer network of the power industry.

Communication protocols are being developed basing on that within the PGU monitor networks. Some remark should be made with regards to the LAN. NetWare 286 is only a tool to develop a dedicated (the single-programme architecture of PC-s makes development of reliable, general-purpose, multiuser network not feasible) networking application. We assume that the electronic mail service enables us to interconnect the LPP-s into an efficient and useful configuration. LPP-o is a working station serving also a function of a gateway to POP-BASE. Therefore, LPP-o is to perform:
    - virtual terminal functions (conversion of LAN protocol to POP-BASE/PGU monitor protocol),
    - transport station/network level functions.

FINAL REMARKS AND CONCLUSION

We have widened this section upon request of our referees who demanded us to clarify the innorative acpects of our approach.

After a very good Polish actor, Jan Świderski (not alive), we distinguish between the new and the modern (innorative): something new must be better than the old to be called modern.

In the paper and elsewhere (eg. Lewoc (1989)) we presented our approach to development of computer networks for power plants. The approach basing mainly on sound and effective software solutions on star networks gave a feasible solution for a ca. 2000 MW power plant. Comparing with the new solutions basing on Proway networks and the TOP protocol, our solution is better with respect to price, reliability development time, etc.

Thus it should be called modern. The above statement needed more than 150 man-years to be proved. And we do hope that this work - a collection of many more or less obvious solutions described here and elsewhere - may be called innovative since it gives a complex and verified base for development of real-time computer systems for the power industry.

The bottom-up method enables to develop a complex but efficient distributed data base for real-time applications in power plants. The approach results in economic solutions and highly decreased chances of major design errors.

The solutions discussed here are to provide organizational bases (standards) for development of distributed data bases for Polish power industry. Similar methods and tools may be implemented in other applications (some projects are in progress).

REFERENCES

Błach L.K., J.B.Lewoc, J.Mertz (1989), Comparison of the top-down and bottom-up strategies in health service computer system development, In Information System, Work and Organization Design, IFIP, Berlin.

Digital (1982), PDP-11 processor handbook, Maynard, Massachussetts.

IASE (1985), Master-2 processor handbook, Wrocław (in Polish).

ISO (1982), Open Systems Interconnection: Basic Reference Model.

Lewoc J.B., K. Misiak, A. Tomczyk (1989), A complex automation and management network for power plants, In Energy Systems, Management and Economics, IFAC, Tokyo.

Meraster (1989), Microcomputers MERA 680, Katowice (in Polish).

Novell (1986), NetWare user reference, Utah, USA, V. 2.00.

Rozent M. and co-workers (1989), Functional (technical) project of CADEL, IASE no. 419 (in Polish).

A REAL-TIME DISTRIBUTED DATABASE SYSTEM IN COLD TANDEM MILL

Jiahua Wang, Huaiyuan Zheng and Jiakeng Zhu

Department of Computer Science and Engineering,
Northeast University of Technology, Shenyang Liaoning, PRC

Abstract. This paper describes requirements of distributed database systems for high speed cold tandem mill systems, analysises properties of transactions in rolling environment using transaction-data model. In order to meet the requirement of prompt response to the process control, concurrent control and reliability provisions being suitable for real-time distributed database systems are represented.

Keywords. Steel industry; rolling mills; real-time computer systems; distributed databases; distributed control; system failure and recovery; concurrent control; globle serializable.

INTRODUCTION

Since computer control systems were applied in the rolling field in 60's, the quantity and quality of products have greatly been raised. A.Daneels and P.Mead (1977) designed a database system for multi-computer real-time control system used in accelator control environment. A real-time distributed database system has been used in naval command and control environment. Recently distributed computer control systems have been used in rolling field. We try to make an effort in applying distributed database approach to rolling control systems, which would be able to improve not only system reliability but also data availability.

RAM memory used as medium of database has been concerned because of its fast response time. D.Dewitt and colleaques (1984) investigated main memory database techniques. In market there have been non-volatile RAM storage board whose access time is magnitude of nano-seconds. They can interface with uni-bus and Q-bus and are inexpensive. Because no more kinds of storage boards can be selected, we use common RAM memory plus backup battery as medium of database.

THE REQUIREMENTS OF ROLLING SYSTEM

Rolling computer control systems are notably different from common commercial computer systems such as bank systems. The response time of commercial computer systems ranging from several seconds to several ten seconds is acceptable. In rolling systems, the requirement of sampling time is between 10 ms and 20 ms. For high-speed roll mill, strip velocity is over 50 m/s, sampling time is shorter so as to raise quality of products. If response time was not satisfied, the results would be dangerous, producing bad products, even damaging devices or injuring people. This system is also different from some real-time systems such as a naval command system. The latter processes data of little volume and volatileness. the system is shut down, all data are not required. For example, the coordinate, velocity and track observed this time are of little important for future use. In case of rolling system data taken this time not only are currently used in rolling process but also are of very important for rolling same steel later. A real-time distributed database system in mill serves correctly process control as well provides data for CAD design of passing plate shape, development of mathemetic model and production management etc., which incleases greatly system complexity.

SYSTEM CONFIGURATION

Our real-time distributed database system is based on a 3-stand cold tandem mill installed in our rolling steel laboratory. Though experimental rolling may be intermittent, our goal is oriented to continuous production of plants. querying process data is performed at the same time as rolling process. This is equivalent to continuous productive environment. This system is shown as Fig. 1.

the system is defined as two levels: process control and experiment management. Experiment management level is consisted of two SUN computers connected by MAP broadband network, process level is consisted of four microcomputers connected by MAP carrier network. The two networks are connected by a bridge. The microcomputers are used for acquiring data and controlling rolling process. One of two SUN machines is used for CAD design, the other is used as experiment management

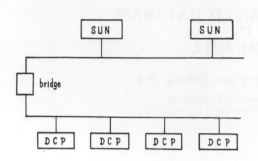

Fig. 1. System configuration

and process control computer. The process control
computer calculates mill setup and traces slabs.
On the management level the database management
system is SUN-INGRES, On the process control
level, is being developed a real-time distributed
database management system which is based on
relational model.

DATA ALLOCATION AND DATA QUERY

In rolling control process, various data are
mutual relevant, for example, mill setup computed
by the control computer is concerned with every
stand. Thus, the mode of data allocation in this
system is fragment and replicating. For example,
mill setup data are stored in the control
computer, the relevant fragment is also stored in
every microcomputer. Each process data item has at
least two copies. This is shown in Fig.2.

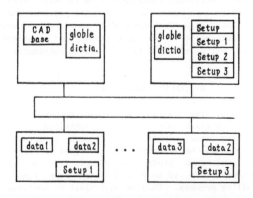

Fig. 2. Data allocation

The users on the management level query data
stored in any database on network using
distributed INGRES without involving their
physical location. The distributed INGRES will be
implemented by way of adding distributed
interface to SUN-INGRES. In order to have INGRES
accept fast process data from procss control
level, a real-time interface has been added to
INGRES.

THE PROMPTNESS CONTROL OF ROLLING PROCESS

Lann (1983) investigated in depth the requirements
of real-time distributed computing including
promptness. We emphatically discuss promptness in
a real-time distributed database system in tandem
mill systems. Rolling process requires sampling
time ranging from 10 ms to 20 ms, or less. During
this period, a number of functions should be
completed, which includes taking data, reading or
writing database and calculating for control
action, etc. In order to meet the strict
requirement of time, besides using RAM memory as
medium of database, we must carefully investigate
the algorithm of transaction syncronization.

The Properties of Rolling Transaction

In commercial distributed data base systems,
concurrent control mechanism usually uses globle
2-phase lock or time stamping mode to obtain
transaction serializable schedule. The two modes
allow transactions keep hold of resources
exclusively for a long time. This is not usually
permitted in real-time systems, therefore we must
find a simpler and more practical syncronization
protocal according to real-time transaction
properties. The transactions in rolling process
are described as follow:
Sensor writing transaction Tsw which writes data
in a specified relation.
Control unit transaction Tc which reads data from
a specified relation, performs calculation and
drives a specified unit.
Setup dispatch transaction Tsd which dispatches
mill setup data to the microcomputer of a relevant
stand according to tracing.
Process data query transaction Tq which queries
process data in behalf of a user on management
level for his/her application.
The associations between the above mentioned
transactions and data are shown as Fig.3. We call
this figure transaction-data model.

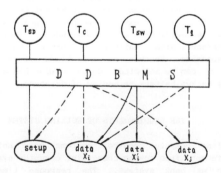

Fig. 3. The transaction-data model.

In Fig.3. solid line indicates writing operation
and dot line reading operation. Two transactions
are in conflict iff they read/write the same data
and at least an operation is writing. Sensor

writing transactions Tswi and Tswj respectively
write own data Xi and Xj. Xi and Xj are disjoint,
thus there are no conflicts among them without
syncronization; Because both control unit
transaction and process data query transaction are
only-reading transactions, there is no the
requirement of sycronization between them; Tsd and
Tsw write disjoint data, they do not conflict with
each other. Among all the above mentioned
transactions, only are in conflict reading
transaction and writing transaction such as Tsw
and Tc, and the conflict only is on a data
relation. furthermore a transaction access a data
relation no more than once. No matter how carried
out, these transactions always are globle
serializable. It is only requirement that mutually
exclusive access be preserved. In the above
situation, transactions Tsd,Tc,Tsw and Tq are in
conflict. It is supposed that they arrive at the
same time. For example, if they are performed in
following sequence: Tsd->Tc->Tsw, Tq->Tsw, there
exists a serial sequence: Tsd->Tc->Tq->Tsw. They
are compatible. Notation -> denotes prior to
relation.

We now prove that these transactions are carried
out in any sequence, they always are globle
serializable. We suppose that there exists a cycle
sequence Tsd->Tc->Tq->Tsw->Tsd. There must be two
transactions, e.g. Tc and Tq, which have
respectively two reading/writing operations. These
operations are interleaved with each other. The
antecedent is that two transactions only are in
conflict on a data item and each transaction only
operates this data item once. there is
contradictory, the preposition is true.

Lock Protocal and Lock Modes

In order to dead lock free, it is specified that a
transaction only acquires one lock at a time,
acquiring a new lock after releasing a old lock.
We prefer preemptive lock to spin lock so as to
avoid wasting CPU time. Lewis (1982) described in
detail the two lock methods. In waiting queue, is
suspended a task which does not obtain a lock,
once the required lock is released, the task is
immediately run. In order to meet the requirement
of promptness, it is necessary that conflict with
each other be avoided, or be as few as possible,
if any. Thus the following should be implemented:
1. The transactions reading the same data can be
concurrently performed.
2. The transactions can be concurrently performed
which read/write different data in a relation.
3. Wait time for access data is limited in
magnitude of tens microseconds.
To implement the above requirements, lock mode,
lock granularity and storage structure must be
carefully arranged. Data of a relation are defined
as history and current records, The last inserted
record is called current record and the others
history records. According to properties of
transaction operations and operands, lock mode is
defined as:
reading history data lock (RH),
reading current data lock (RC),

writing current data lock (WC) and
deleting history data lock (DH).
By means of the four lock modes, the before-
mentioned requirements 1. and 2. may be met. With
this arrangement, lock table is small in size,
thus overhead of lock and unlock operations is
little. This is of great benefit to decreasing
wait time.

Storage Structure

The storage structure of relations is arranged as
a loop list. We call a relation implemented in
RAM memory memory file. To minimize access time,
we place two pointers: head pointer and rear
pointer in every memory file. The former points
the first record of a relation and the latter last
record, or current record. This is shown in Fig.4.

lock table

tran.ID	data.ID	mode
T_q	data 1	RH
T_{sw}	data 1	WC

data 1

head ptr.
rear ptr.
history data
Current rec.

Fig. 4. Storage structure.

Access current record is carried out by means of
rear pointer. When a sensor writing transaction is
writing a new current record, the reading
operation of a control unit transaction is
delayed. This delay can be limited in magnitude of
tens microseconds. When history records are
deleted or dumped, free space can be fast
collected by way of changing head pointer without
moving rest data.

MUTUAL CONSISTENCY OF COPIES
AND FAILURE RECOVERY

To obtain high system reliability and query
localization, the mode of data allocation in this
system is fragment and replicating. This results
in a problem: how does copies' mutual consistency
be preserved? Mutual consistency of copies means
whenever copies of the same data are
simultaneously read by transactions in differnt
sites, the results should be the same. To address
mutual consistency of multi-copies, the
investigators in the database field
(Berstein,1980,1987; Ceri,1984; Holler,1981) have
represented a number of algorithms. To implement
these algorithms, it is necessary that sessions be
frequently set up between sites. The communicating
protocal for real-time control is usually token
passing, it is impossible that sessions are
frequently set up during tens milliseconds. Under
support of reliable communicating network , our
system follows so-called principle of reliable
syncronization writing. It is based on the
following assumptions:
1. Reliable communicating network deliveries

exactly messages.

2. Reliable buffers do not lose any information.

3. Data update is only carried out by owner or its agents.

The algorithm is described as follow:

1. When accepted by the owner of master copy, data to write are immediately written into its reliable buffer and broadcasted to relevant agents.

2. While received by agents in receiver sites, data are written into agents' reliable buffers.

3. When the communication is over, the owner and its agents in receiver sites set up relevant locks, write data from reliable buffer into specified relation and verify correctness. No further communication is needed.

4. The owner and its agents release their locks.

By means of buffering and communicating, data copies are syncronizably updated, the mutual consistency of multi-copies is preserved.

Recovery of failure is simple, because data are replicated, every data item has at least two copies which are dynamically maintained. Thus, the failure of any site can be recovered by relevant copies.

CONCLUSION

Although this system is developed for high-speed tandem roll mill, it will be suitable to other real-time control systems which have transactions of the same properties. The data access method of this system is implemented by way of pointers. Query of history data is sequentially performed. It is impossible that a specified history record is randomly retrieved. A new access method of dynamic hashing with preserving sequence has been developed for more complicated control systems.

REFERENCES

Bernstein,P.A., D.W.Shipman,and J.B.Rothnie,JR. (1980). Concurrent control in a system for distributed databases (SDD-1). ACM Trans. database system. vol5. No.1. March, 1980. 18-51.

Bernstein,P.A., V.Hadzilacos, and N.Goodman (1987). Concurrency and control recovery in dadabase systems. Addison-Wesley Publishing Company.

Ceri,S. and G.Pelagatti (1984). Distributed Databases Principles and Systems. McGraw-Hill Book Company.

Daneels,A. and P.Mead (1977). Impletmentation of a database in a distributed computer system for real time accelarator control. International conference on distributed computer control systems 26-28 September 1977.

Dewitt,D., AND Others (1984). Implementation techniques for main memory database systems. In B.Yormark (Ed.) SIGMOD'84 Proceedings of annual meeting vol.14. No.2.

Holler,E. (1981). Multiple copy update. In B.W.Lampson and others (Ed.) Distributed Systems-Architecture and Implementation Springer-Verlag Berlin Heidelberg New York pp284-307.

Lann,C.L. (1983). On real-time distributed computing. In R.E.A.Mason (Ed.) Information Processing Elsvier Science Publisher B.V.(North-Holland).

Lewis,T.G. (1982). Software Engineering Analysis & Verification Reston Publishing Company, Inc. A Prentice-hall Company, Reston, Virginia. 284-327.

AUTHOR INDEX

KEYWORD INDEX

Printed and bound by CPI Group (UK) Ltd, Croydon, CR0 4YY

03/10/2024

01040320-0015